Stars and Space with MATLAB Apps

With Companion Media Pack

Other books in this series by the author

One Hundred Physics Visualizations Using MATLAB (2013)

More Physics with MATLAB (2015)

Cosmology with MATLAB (2016)

Beams and Accelerators with MATLAB (2018)

Stars and Space with MATLAB Apps

Stars and Space with MATLAB Apps

With Companion Media Pack

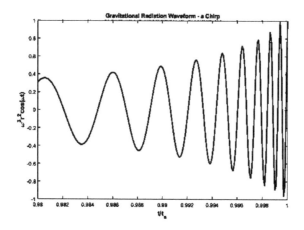

Dan Green

Fermi National Accelerator Laboratory, USA

 **World Scientific**

NEW JERSEY • LONDON • SINGAPORE • BEIJING • SHANGHAI • HONG KONG • TAIPEI • CHENNAI • TOKYO

Published by

World Scientific Publishing Co. Pte. Ltd.

5 Toh Tuck Link, Singapore 596224

USA office: 27 Warren Street, Suite 401-402, Hackensack, NJ 07601

UK office: 57 Shelton Street, Covent Garden, London WC2H 9HE

British Library Cataloguing-in-Publication Data
A catalogue record for this book is available from the British Library.

STARS AND SPACE WITH MATLAB APPS
With Companion Media Pack

ISBN 978-981-121-602-2 (hardcover)
ISBN 978-981-121-635-0 (paperback)
ISBN 978-981-121-603-9 (ebook for institutions)
ISBN 978-981-121-604-6 (ebook for individuals)

For any available supplementary material, please visit
https://www.worldscientific.com/worldscibooks/10.1142/11707#t=suppl

"Look up at the stars and not down at your feet. Try to make sense of what you see, and wonder about what makes the universe exist. Be curious."

— **Stephen Hawking**

"Just as the system of the Sun, planets and comets is put in motion by the forces of gravity, and its parts persist in their motions, so the smaller systems of bodies also seem to be set in motion by other forces and their particles to be variously moved in relation to each other."

— **Isaac Newton**

Preface

"Finally we shall place the Sun himself at the center of the Universe."

— **Nicolaus Copernicus**

"Humanity has the stars in its future, and that future is too important to be lost under the burden of juvenile folly and ignorant superstition."

— **Isaac Asimov**

We are living in a time of great advances in physics and astrophysics, both scientific and technical. The last fundamental particle in the Standard Model of particle physics, the Higgs boson, was discovered in 2012 and the Nobel Prize awarded the following year. Our knowledge of the basic forces and particles seems secure at present, although many questions still remain to be explored.

Cosmology, using new technical tools, has become a quantitative science albeit by invoking new forms — dark energy and dark matter. Both dark entities have no known candidates in the present knowledge of physics, but their direct observation is now being very actively attempted. The Standard Model of cosmology explains the basic large-scale structure and evolution of the Universe in terms of a handful of basic parameters.

The Big Bang fireball was an ionized plasma of photons, with a small admixture of electrons and nucleons. As the plasma expanded and cooled, nuclei, largely hydrogen and helium, were formed. The existence of a matter asymmetry is not understood. Nevertheless, assuming a predominance of matter, the neutrons were all stably and deeply bound into helium. Other nuclei existed only in trace amounts in the "primordial" composition of matter. Attempts to use fusion on Earth as a source of clean power continue with the massive effort to build and operate the International Thermonuclear Experimental Reactor (ITER) in France and in basic research on other methods to initiate fusion.

Quantum fluctuations in the plasma evolved into large scale structure — galaxies and stars — since gravity is always attractive. The twenty-first century tools of astronomy were used to make surveys of galaxies using visible, infrared, radio and X-rays using instruments that are both Earth and space based. Very recently, the direct observation of gravitational radiation has added a new observational tool in the exploration of the Universe and resulted in the reward of a Nobel Prize in 2017. This new technique is analogous to the transition from Coulomb static to electromagnetic wave dynamics.

Supernovae have been seen, and the spectral analysis shows that they are the source of heavy elements. Indeed, the Sun has elements in it heavier than iron, which indicates that the Sun is at least a second-generation star since elements heavier than iron cannot be formed by normal fusion processes of light elements. White dwarf stars have been surveyed, and pulsars are the visible manifestation of neutron stars. In 2019, a picture of a black hole event horizon explicitly indicated the existence of an event horizon as predicted by General Relativity (GR).

It is just 50 years since the first human landed on the Moon. On a planetary scale, landers are on the surface of Mars. The moons of Jupiter have water ice. Deep-space probes have escaped the Sun to travel into interstellar space. Astronomers have found many exoplanets orbiting distant stars, some rather Earth-like. Attempts to determine the atmospheres of these planets are ongoing. It is now known that planets are common and that Earth-like planets exist.

In the realm of human exploration, new rockets with reusable booster stages are now operating. That feature promises to make near-Earth orbit operation much cheaper than the single-use boosters that were used in the past. Prototype solar sails are in existence, and the use of launch lasers might make interstellar exploration possible using sails with a laser assist — if slow. Initial prototypes of ion thrust rockets that greatly increase the exhaust velocity of rockets are active.

All of these new developments inspire an attempt to use new Matlab tools to explore some of these topics. The use of "Apps" is a chance to put "old wine into new bottles". The aim of this book is to make the topics covered in this book available using a single App tool. The App tool is an ideal teaching method as it incorporates within a single Matlab Figure plain text and formulae while creating numerical and graphical output with input to be varied by the user using a variety of methods. This tool unifies a problem into a single Matlab Figure which connects input to output with any changes in the inputs immediately evident in the outputs.

Introduction

The specific aim of this book is to show how the physics of stars and space travel depends on the parameters of the specific problem being addressed. It does not aspire to teach the basic physics but simply to explore the ramifications of the physics. That being said, details of the formulae quoted are not simply stated but are motivated by basic physics intuition. When the physics remains obscure, recourse to the web, either a general browser search or a specific Wikipedia query should be a sufficient aid to supply more detailed physics context. In fact, there are no print references supplied in this book. A modern variant to a long list of inaccessible hardcopy references is a search through the internet. Indeed, in the dawning age of open-source material available to anyone, this seems to be the appropriate future direction when one wants to access specific knowledge.

Some knowledge of classical mechanics (CM), electromagnetism (EM), quantum mechanics (QM), and special relativity (SR) and general relativity (GR) is assumed in this book. Nevertheless, this book provides the basic equations to be explored and a sketch of their derivation while also providing an explanation of how the equations at issue have been derived. The approach of this book is that a picture is worth a thousand words, especially if the picture changes in real time as the user changes the relevant parameters that specify the particular solution of a problem.

The tool used for exploration is the Matlab — software package. Matlab provides a complete ensemble of tools to attack almost any problem in physics. In addition, the package provides many tutorials and many useful search features. Indeed, many colleges and universities have site-wide licenses so that the package would be freely available to students. If such facilities are not available to the reader, an inexpensive student version of Matlab is also available. A first-time user has an available introductory tutorial to start off. There are many more specialized tutorials that can be usefully studied. Finally, there is a help facility that searches the entire Matlab documentation for a relevant response to a query. Matlab is also compatible with the Python language, so that users can mix scripts if that is wished.

Specifically, the focus of this book is on stars and space travel. Chapter 1 is a short introductory, containing a brief introduction to Matlab tools. This chapter is quite superficial because Matlab provides such an extensive suite of tutorials and examples. The distinction between scripts, .m files, and Live scripts, or .mlx files is made using example problems.

Chapter 2 focuses on the "App" tool which provides the input and output on a single Figure. In that way the use of command-line queries with distinct printed output and separate distinct figures in scripts is unified in the App. In this way, the App tool is so appealing that almost all the topics explored in this book are accomplished using the App tool. The code for this chapter is provided in Appendix A, which provides a few examples of the App Code View. Those examples are supplied to give a first-time App user a view of the structure of Matlab code in simple cases.

After this introduction to the App tool, Chapter 3 is about formation and characterization of stars. Chapter 4 is about stellar evolution toward the end points of white dwarf stars or neutron stars or black holes. Which branch is the endpoint of the star depends on the details but largely is determined by the mass of the star. The discussion then shifts from stars to planets and orbits and the exploration of the local solar system.

Chapter 4 is about planetary orbits, rocket operation, and planetary exploration. The focus is on launching from Earth, docking in Earth orbit, landing on other solar bodies, and finally returning to Earth through the atmosphere. Some other possible methods of planetary travel beyond just rocketry are also mentioned.

Chapter 5 discusses the possibilities of interstellar exploration. However, only established physics is invoked, and the somewhat depressing consequences are faced. Travel between the stars will be slow. Relativistic effects are examined since extreme speeds are at a premium over distance of light years. If wormholes exist, things would be a lot easier, but this book is limited to well-understood physics.

Finally, Chapter 6 showcases some applications of Matlab to examples in EM, QM, sounds using Fourier transforms, and cosmology. These examples serve to illustrate a few of the Matlab utilities unused in the main sections of this book, such as partial differential equation tools.

Appendix A explicitly shows the scripts for the examples explored in Chapter 1. All the scripts are made available by the publisher at https://www.worldscientific.com/worldscibooks/10.1142/ 11707#t=suppl In Appendix B a list of the symbols used in this book are provided. Appendix C tabulates some basic properties of the Sun, while Appendix D gives some numerical properties of the solar planets. Some properties of the Earth's atmosphere are tabulated in Appendix E. Finally, the acronyms used in this book are tabulated in Appendix F.

Contents

Chapter 3. Stellar Evolution   61

Chapter 4. Apps for Solar Exploration   103

Chapter 1

Getting Started with Matlab

"Access to computers and the Internet has become a basic need for education in our society."

— **Kent Conrad**

"But the most remarkable discovery in all of astronomy is that the stars are made of atoms of the same kind as those on the Earth."

— **Robert Feynman**

1.1. Windows and Editor

The Matlab product is available for users at many universities and businesses. When Matlab is first executed, the Command Window is opened. The ≫ prompt indicates Matlab is ready for a command. Figure 1.1 shows the Command Window just after invoking Matlab. Added displays of the command history and the workspace variables are useful and can be docked at the periphery of the Command Window using the "Home" tab with "Layout" selected in the dropdown menu.

Using the "Search Documentation" window on the top right, first search for "getting started". This opens the Matlab Documentation, which has many choices. Choosing getting started with Matlab has many possible topics and tutorials. Detailed responses to queries can be obtained using the "Search Documentation" window at any time.

The "Search Documentation" allows the user to search the extensive set of Matlab topics and tutorials. For example, a search on "function" opens up many explanations of the available functions.

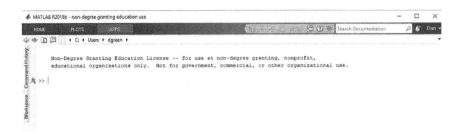

Figure 1.1: Command window appearance at the start of Matlab. The "Workspace" window and the "Command History" window have previously been added to the layout and docked.

A search on "symbolic math" enables an explanation of the available functions and solvers and supplies an extensive tutorial for symbolic math tools using Live script.

The "Home" tab has an "Open" tab to open editors or create new scripts. The scripts supplied by this text can then be invoked and run. One first has to use the "set path" button to indicate where your scripts can be found or where user data are stored.

In this fashion, the new Matlab user can quickly search for help and quickly build up her expertise with the very extensive suite of Matlab capabilities. In future examples and Apps, new tools that are used in the scripts supplied in this text will be explained as they arise, case by case.

The "HOME" tab expands for the user as displayed in Figure 1.2. The major use of this tab in this text is to create new scripts and Apps, using the "New" or "Open" tab and to use the "Editor" this invokes to write, save, open, and debug scripts in the "Editor" window. In all cases in this text, scripts will be opened from the "Command Window".

Figure 1.2: The drop down "home" tab in the "Command Window". New scripts can be created or existing scripts or Apps can be opened in the "Editor" window.

Using the "Open" tab, the user will invoke the "Editor Window", for an existing script when the tab lists scripts which are on the path set initially. Choosing "Editor_Script.m" opens the script in the "Editor" window as shown in Figure 1.3. The example shown in Figure 1.3 shows the color coding of the instructions. The main tab is the "Editor" with options which are largely self-explanatory.

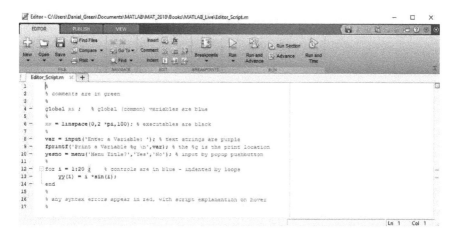

Figure 1.3: The result of choosing "Open" in the "Command Window" and then choosing the script "Editor_Script.m".

During script creation, using the "New" tab, the Editor indicates incorrect statement lines and what the issue might be — in red. Hovering on the name of a utility brings up a short indication of the required syntax, Comments, with a leading % character, are in green. Global variables are shown in teal. Printout in the "Command Window" is suppressed by ending a line in ";". Input variables are entered in the "Command Window" by using "input" to ask the user for command line input. There is also a popup "menu" utility which allows the user to click on specified choices. Output variables are printed in the "Command Window" using the command "fprintf". Executable operations are shown in black, while text strings appear as purple characters. Clearly, this is a lot of material to take in for a new user, so several simple examples are given to ease into the first use of Matlab.

1.2. Script and Live Coding Options

Simple problems in Matlab can be done entirely in the Command Window. More complex problems, but still rather simple, can be done by writing a script in the "Editor" by asking for a "New"/"script" in the "Command Window" which invokes the "Editor" where the new script can be written, named, and saved. As a tool to explore physics, this method is somewhat limited. The script is seen in the "Editor", the output and input is handled in the "Command Window", and the graphical results, if any, appear as a distinct Figure in yet a third window. This sequence has a tendency to require the user to juggle three windows at least.

Because of this cumbersome division between scripts, input and output and graphical plots, Matlab has evolved, over time, tools to make Matlab a more integrated experience. Indeed, this text uses the "App" tools almost exclusively because they allow input, output, and plots to all appear on one Figure and the problem variables to be changed in that Figure with immediate results displayed in figures or numbers on the same global Figure.

An illustrative example of the possible ways to find and plot the Taylor series of a function is explored below using the script "Taylor_script". The scripts use symbolic variables and symbolic functions and the "plot" utility is used to make plots of the input function and the output polynomial Taylor series approximation to the function.

Expanding about $x = 0$, the Taylor series for a function $f(x)$ is given as follows:

$$f(x) = f(0) + x(df/dx)_o + (x^2/2)(d^2f/dx^2)_o$$
$$\cdots [x^{n-1}/(n-1)!](d^{n-1}f/dx^{n-1})_o \qquad (1.1)$$

The first step is to search the Matlab documentation for a utility to find the Taylor series. Figure 1.4 shows the result of a search for "Taylor" entered into the "Search Documentation" window. In this case, x is defined to be a symbolic variable. Thus, f is also a variable. The eighth-order Taylor expansion is printed because the (;) at the end of the line which is normally used to suppress printing is deleted.

Taylor Series

The statements

```
syms x
f = 1/(5 + 4*cos(x));
T = taylor(f, 'Order', 8)
```

return

```
T =
(49*x^6)/131220 + (5*x^4)/1458 + (2*x^2)/81 + 1/9
```

which is all the terms up to, but not including, order eight in the Taylor series for $f(x)$:

$$\sum_{n=0}^{\infty} (x-a)^n \frac{f^{(n)}(a)}{n!}.$$

Technically, T is a Maclaurin series, since its expansion point is a = 0.

Figure 1.4: The result of a query to the "Search Documentation" window for "Taylor". In most cases, examples are given along with the basic syntax. The command lines entered into the "Command Window" are shown outlined in gray.

More tools are used in the script "Taylor_script.m". The script, identified by an .m designation, is useful for simple tasks. The Matlab utilities "syms", "simplify", "pretty", "menu", "fprintf", "taylor", "figure", and "plot" are used in the script, but the output is not shown here. The interested user is encouraged to run the script and explore the options which are made available by the script.

The "Live" option is more optimized for formal teaching. There is an equation tool to add equations to the output as well as general textual comments. They are invoked with the "insert" tab in the Live editor with choices of "section break", "text", and "equation" among other possibilities. There is documentation available to the user in the tutorials "Teach with Live Scripts" and "Introduction to Live Editor" in the Matlab documentation search window.

The "New" tab allows the user to create a new Live script by choosing "Live script", which opens the "Live_Editor" window. Using the "Open" tab in the "Command Window" "Home" tab allows the user to search for files — script, Live, or App. Choosing a Live file

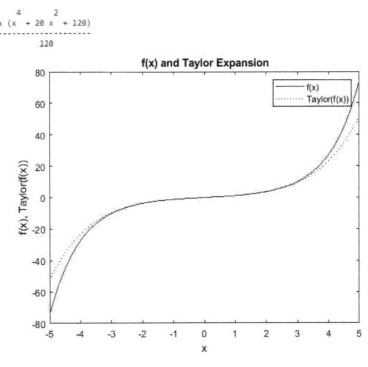

Figure 1.5: Output of the Live code for the Taylor series example. The symbolic solution appears followed by the plot of the function itself and the approximate Taylor series to order 8.

invokes the "Live Editor". Using "Run" in the editor executes the Live script, identified by the .mlx tag.

Output of the live script, "Taylor_Live" appears in the Live editor (Figure 1.5). The code and the output can be viewed together in line with a white background for output and gray for code. Section breaks can be used to evaluate the output at each step which aids in the visualization connecting the input code and the output at multiple locations in the code. The series appears on the screen directly followed by the figure which is an improvement over the script example since the text and figure output both appear in the same screen. However, user input is still entered in the command line dialogue and control uses the "Menu" popup as in the case of the script. In this text, the "Live" script option will not be used, because the simple .m script works well for small problems while the "App" option puts all aspects of the problem on a single Figure canvas.

1.3. Using the App Coding Tools

The user may type a few simple command line instructions for easy explorations. However, some saved script is most often desired for later use and modification and improvement. The .m type script has a distinct "Editor" window, with the "Command Window" used for input and output and a distinct "Figure" window needed for each separate plot. For the "Live" coding option, the code and numeric output are available in the same window and the plots can be simultaneously viewed either in line or via a split screen. However, changes of variables for a problem still are made via the Command window. The experience for the user is more unified if the "App" option is used.

To get started, a search from the Command Window on "Search Documentation" for "App Designer" yields many suggestions. Picking the first suggested topic supplies the user with a tutorial with five distinct examples. These examples are a good place to start. A distinct "Editor" is invoked for the "App" files which have the ".mlapp" identifier.

Using the "Command Window" tab for "Home" choosing "New" and then "App" invokes the "App_Designer". That window offers a tutorial, four simple examples and seven more complex examples to explore. After that the user can choose a Figure to create the App as desired. All of the Apps used throughout the text have parameters set initially in the code so that when the "Start" button is pushed, the user sees a valid solution to the problem. Then the user can change some of the parameters and immediately see how the solution changes.

In order to first compare App to Live and script the Taylor series is redone as an App. The "Command Window" tab "Open" with the example "Taylor_App2" chosen opens the App in the "Design View" as seen in Figure 1.6. A Text box, Start button, and numeric Edit Fields define the input problem. The output series appears in Text

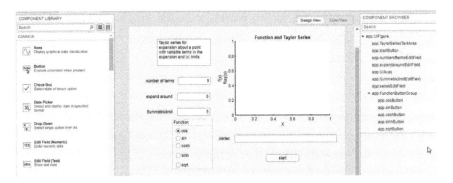

Figure 1.6: The "Design View" of the "Taylor_App" script. The component library also appears on the left, while the browzer for the chosen components appears on the right.

Edit Field and is plotted using the Axes component. The names of these components appear in the Component Library. The interested user is encouraged to run the App and compare to the script and "Live" versions of the same problem in order to judge the superiority of the App formulation. A specific output of the App will be shown later, Figure 1.15, after a few more App examples are shown.

The "Editor" window is specific for the "App Designer". The view can be toggled between "Design View" and "Code View" (Figure 1.7). The code itself, along with the code for all examples in Chapter 1, is shown in Appendix A. The Editor tab allows the user to create or edit the code, save it, or Run it. The "callbacks" refer to code to respond to changes in the values of the inputs. The color coding is the same as the script Editor. Several distinct functions can be added to the code as desired using the add Function button. The actual code is taken from a tutorial example provided by using "App Designer". Besides the code itself, windows for the code browser, component browser, and the design view layout are available.

In this first example, the Design View is as simple as an "Axes" plot and a "Slider" defines the function to be plotted, which is a Matlab utility function called "peaks". When the Slider is changed, a "callback" is invoked in the Code View. The Slider value is used to set the function amplitude. User supplied code is in white while the boilerplate App code is in gray and should not be altered.

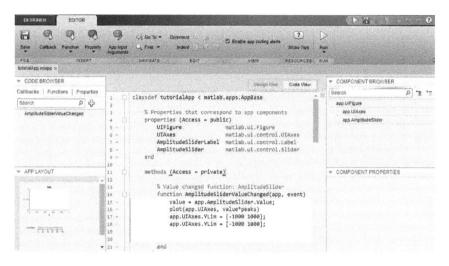

Figure 1.7: The Editor screen in the Code View of the App Designer. The user code has a white background, while the fixed code specific for the App Designer is gray and is not to be changed by the user.

1.4. App Examples

A few simple App examples are shown first before using almost always Apps for the main topics of stars and space exploration in the text. They should serve as an introduction to simple use of the App tools. The reader can Run them and look at the code in the Code View or examine the code in Appendix A. In general, a "Start" pushbutton makes a complete code execution. Then the user changing parameters update the execution so that the effects can be seen in real time.

App Designer supports most MATLAB graphics. Initially, only "plot" was supported, but as of the 2018 release, almost all graphics are usable. Exceptions are rare, such as "cga", but they can be found in the Matlab documentation.

The "Plot_Demo" App shows a simple plot using the "plot" utility (Figure 1.8). There are many possible App components to choose from. In this text, the choices are mostly limited to Text box, Axes (plots), Buttons (start), Edit Fields (Text and Numeric), menu buttons, and Sliders (change parameters). The "Start" button

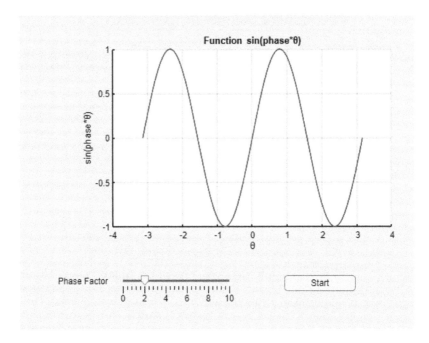

Figure 1.8:   Figure created by "Plot_Demo". After starting the display with the Start button the Slider can be used to alter the phase factor of the sine wave.

makes the initial plot. In this case, a change in the "Phase Factor" Slider generates a callback which causes the plot to be updated. The code uses the Matlab utilities, "plot", "title", "xlabel;", "ylabel", and "grid".

A few of the Matlab three-dimensional graphics options are explored in the script "Demo3d_App". A particular output is shown in Figure 1.9. Plot options for the same function can be chosen using the utilities "mesh", "meshc", "surf", and "surfc", where the ending c implies the creation of contour plots. The type of plot is chosen by the user clicking on the "radio" button. In this case in the Code View, a function called "updateplot (app)" is added to be invoked first with the start button or later by a change in the type of plot. The Matlab utilities used here are "meshgrid", "colormap", "Xlim", "Ylim", and "Zlim".

The script "Data_Examine_App" uses Monte Carlo techniques to generate model masses distributed as a resonance line shape,

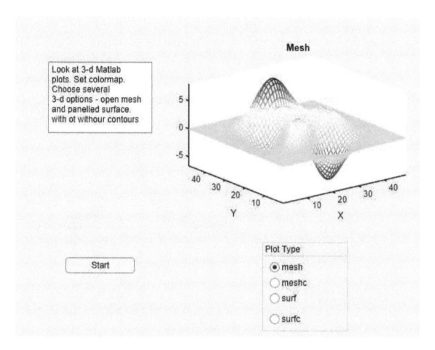

Figure 1.9: Figure for the App "Demo3d_App". The same function can be plotted in four distinct ways.

a Lorentzian. This shape is defined by a peak mass, M_o and a full width at half maximum, FWHM, Γ, as defined in Eq. (1.2). The data are generated using the Matlab random number generator "rand" and histogramed using the Matlab utility "hist". The data mean and standard deviations are computed using the utilities "mean" and "std" and displayed using numeric EditFields. The statistical errors are taken to be the square root of the number of events in the histogram bin and are displayed using the plot utility "errorbar". Other Matlab utilities used in the code are "tan", "atan", "linspace", and "sum".

$$dN/dM = (\Gamma/2)^2/[(M - M_o)^2 + (\Gamma/2)^2] \qquad (1.2)$$

The user can very clearly show how the histogram smooths out as the number of total events is increased and can also see how the sample mean and standard deviation approach the true values when more events are generated since the fractional statistical errors are then reduced. A specific figure from the App is shown in Figure 1.10.

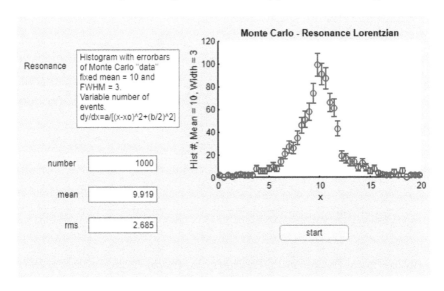

Figure 1.10:   A plot of Monte Carlo "data" for 1,000 events generated with a mean of 10 and a FWHM of 3 where the sample mean and standard deviation are displayed numerically. The rms is not the same as the FWHM, but is close in value.

The equations for a fit of data points with errors to a straight line are provided in the script shown in Appendix A. The "chisquared" function, Eq. (1.3) is a measure of the deviation of the points from the straight line. It is minimized with respect to the variables m and b of the straight line with contributions to the overall χ^2 weighted by the errors, σ_i on the n individual points,

$$\chi^2 = \sum_i^n [y_i - (mx_i + b)]^2 / \sigma_i^2 \qquad (1.3)$$

The App called "Least_SQ_StLine2_App" is invoked to fit some data to a straight line. The algebra to minimize the χ^2 by varying m and b is shown in the code. The code has two added functions in this case, "update" and "leastsq". In general, any number of functions can be added to an App code. By varying the percentage errors on the points, the user can see when the $\chi^2/$dof is about 1. The number of degrees of freedom, dof, is the number of data points, 9, less the number of free parameters, 2, so that dof is 7. Errors of 1% are clearly underestimated, while those of 50% are overestimated and errors of 20% are approximately estimated correctly. Without a correct error estimation, neither the best fit values of m and b which are calculated nor the estimated errors dm and db are to be trusted. The code uses the utility "hold" to allow for the addition of the straight line, shown in red, to the plot of the existing points and their errors. A particular plot is shown in Figure 1.11. If more complex, or nonlinear, problems occur, the user can apply the Matlab fitting tool called "fminsearch".

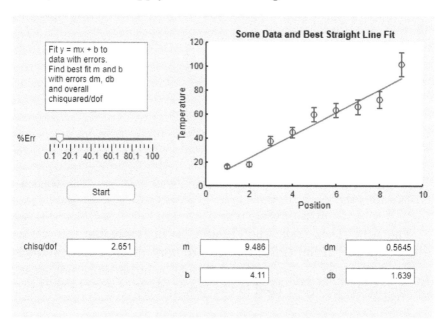

Figure 1.11: Straight line fits to points with errors. The slope and intercepts and their associated errors as well as the goodness of fit are displayed.

Symbolic math is also supported for Apps. In fact, the symbolic solution for many problems can be used so that the user need never look up an integral, a derivative, or the solution to an ordinary set of equations or differential equations. Examples of symbolic operations are differentiation — "diff", factorization, eigenfunctions,

integration — "int", solving algebraic — "solve" and differential —
"dsolve" equations, Taylor series — "taylor", and series summation.
Symbolic variables are declared using the "syms" statement and
appear in the editor colored purple as are text strings, as seen in
Figure 1.12. Some command line examples are shown in Figure 1.12.
In the Apps written for the main body of the text, symbolic math is
used to the extent possible and the symbolic results for solutions are
written into Edit Fields — Text since symbolic variables are treated
as text strings in Matlab.

```
>> syms x f
>> f = sin(x);
>> int(f)

ans =

-cos(x)

>> diff(f)

ans =

cos(x)
```

Figure 1.12:   Command line symbolic math, simple examples. The symbolic
variables are x and f.

The App "Demo_ode_App" is used to illustrate the use of the
utility "ode45" to numerically solve ordinary differential equations
in the case where a closed-form symbolic solution does not exist.
The first-order equation used in this example is

$$dy/dt = a\cos(y)e^y \tag{1.4}$$

In this example, the initial condition $y(0) = 1$ is specified. The
App uses the "global" utility to communicate between functions.
The argument list of the functions could also be used, but in this
text, the "global" tool is almost always used for consistency. The
"ode45" utility defaults to t as the independent variable with y as
the dependent variable. Examples are easily found using the "Search
Documentation" window. The differential equation is referenced as

"ode45(@app.name, t)" referring to the "function $dydt$ = name(app, t, y)". This treatment of the utility "ode45" in an App will be used in almost all the numerical differential equation applications that follow.

A newer topic is the use of Matlab tools to solve boundary value problems, BVPs. A simple example is shown here in order to introduce the technique. Later in the text the use of the Matlab BVP solver enables the user to explore a solution for the structure of the Sun treating it as a BVP rather than an initial value problem, which is not a good representation for solar structure.

A different problem for a differential equation is the use not of initial conditions, as in "ode45", but boundary values, or specified function values at fixed points. These problems are addressed in the Matlab utility "bvp4c". The critical input, a function "solinit", is a guess at the solution. Other inputs are the boundary conditions, specified as residuals and the differential equation itself. This utility is a bit slow and success depends crucially on specifying a good approximate initial solution. Iterations may be needed in general. Nevertheless, this Matlab utility supplies the tools needed to address the boundary value problem, should one arise.

A simple example solving a nonlinear differential equation is treated in the App "BVP_Rad_App" (Figure 1.13) with fixed values of y at $x = 0$ and $x = 1$,

$$d^2y/dt^2 = ay/(1 + by) \qquad (1.5)$$

The solutions depend on the two parameters a and b with two boundary values rather than the two initial conditions in the case of an ordinary second-order differential equation. The equation represents a situation where oxygen is diffusing into a solid, while inside the solid the oxygen is being used up in a chemical reaction. If the diffusion is rapid, the oxygen penetrates deeply into the slab. If not, it is used up so rapidly that there is little oxygen in the interior. The plot shows the oxygen concentration for several different balances between diffusion and reaction rate.

As in "ode45", a second-order differential equation is solved by defining it as two first-order equations, one for y and the second for dy/dx (Figure 1.14). The differential equation and boundary

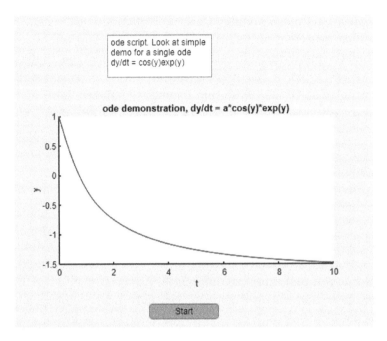

ode script. Look at simple
demo for a single ode
dy/dt = cos(y)exp(y)

ode demonstration, dy/dt = a*cos(y)*exp(y)

Start

Figure 1.13:   Numerical solution of an ordinary differential equation of first order.

conditions are referenced by functions in the main function call as solinit = bvpinit, bvp4c(@app.DEname,@app.BCname,solinit. . . .) which define the equation, the boundary conditions, and the initial starting value for the solution. The plots in this example are overlayed so that the progressive oxygen penetration can be seen. If the user wishes to have only one solution displayed at a time, the "hold" command can be deleted in the code.

The use of Taylor series is finally given as an introductory app in "Taylor_App"(Figure 1.15). The advantages of the App utility can be seen by comparing the previously shown scripts and Live code which addressed the same problem.

All the variable inputs are available in a single Figure. There is an explanatory text box and a "start" button. The series is defined by the number of terms, the function itself, and the expansion point. These can be changed by the use of the EditFields and the RadioButtons and the results appear immediately. The results are

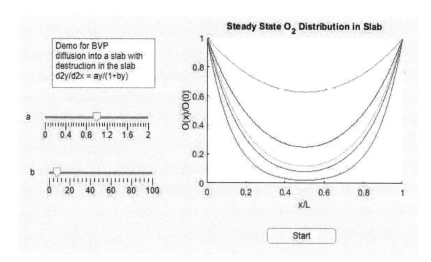

Figure 1.14: Solution of a nonlinear differential equation with two boundary values.

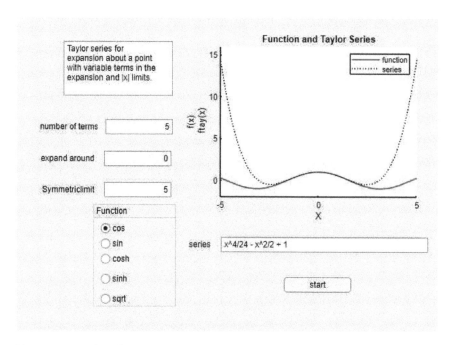

Figure 1.15: App for Taylor series. The user can vary the function, the number of terms in the series, the expansion point, and the limit of the graphical plot. Comparison to the previous script and Live versions is instructive.

plotted over x limits that are also selectable. Finally, the symbolic solution is displayed in a text window. Symbolic variables are used and the symbolic solution appears using the "char()" command to display symbolic variables seen as text strings. The Matlab utilities for symbolic variable "taylor", "simplify", "pretty", and "eval" are used in this App. This compact App structure provides maximum insight and immediacy to the user.

This very brief introduction, with a few examples and with the App scripts themselves given in Appendix A, is assumed to be sufficient to jump into a study of stars, followed by planetary orbits. Then rocket travel is explored to see how space travel is accomplished. Finally, the possibilities for travel to other stars are explored. In what follows, Matlab Apps are used because of their ability to put the entire problem and its solution in a single place and in the hands of the user to explore the effect of parameter variation. In a few cases, very simple examples are given as simple scripts, or .m files. Otherwise, Apps are used in almost all the topics that follow.

Chapter 2

Apps for Stars

"There is stardust in your veins. We are literally, ultimately children of the stars."

— **Jocelyn Bell Burnell**

"It shows you exactly how a star is formed; nothing else can be so pretty! A cluster of vapor, the cream of the Milky Way, a sort of celestial cheese, churned into light."

— **Benjamin Disraeli**

"I think there's something really poetic about using nuclear power to propel us to the stars, because the stars are giant fusion reactors. They're giant nuclear cauldrons in the sky."

— **Taylor Wilson**

2.1. Gravitational Clumping

A system of photons and massive particles can be stabilized by radiation pressure. This stability fails when the system expands and cools. At some point, the gravitational attraction of the massive particles cannot be stabilized by pressure waves in the medium because they have a finite sound speed. Matter and radiation from the Big Bang decouple when the photons can no longer keep hydrogen atoms from forming due to falling temperatures of the plasma.

After the matter and radiation decouple, matter is able to start to form structures. Gravity makes these structures unstable, and they begin to aggregate. The first basic script used is "Grav_Clump_App". It is quite simplified. Particles are put in a box with random initial velocities and positions. Each particle has pairwise attractions with all other particles with a force of magnitude $m_i m_j / r_{1j}^2$ directed along the line of centers. The time is divided into small steps with a change in velocity due to the forces. If two objects get close, they "merge" into an object that is the sum of the masses, $m_i + m_j$, and one mass is dropped or "frozen". This model will ultimately have all the particles clumped into a single aggregate if the system is tracked for a long enough time.

The user chooses the number of starting particles, N, and their velocity scale. The paths are followed and the final number of particles is displayed along with the "mass", which is N divided by the number of particles still unfrozen.

This very schematic demonstration serves to show that low-density situations tend not to clump much, while high-density ones do, yielding high masses. The role of the velocity is supposed to mimic the effect of pressure, but the fact that the masses are constrained to remain in the box (specular reflections) masks the effects of pressure somewhat. Nevertheless, this simple model gives some insight into gravitational clumping and the restraining effects of low density and high pressure, represented here by velocity.

A particular scenario is shown in Figure 2.1. The user is encouraged to play with this toy model and, perhaps, gain some intuition. The density effect, N, is quite strong. The Matlab utilities "floor" (round N to integer), "rand" (random number), and logical operations "for", "if", "while", and "end" (control nested loops) are used heavily in the App. Better representations of the clumping behavior will follow with Apps presented later.

2.2. Neutrons and Protons

How were stars created? The answer requires exploration of the early moments of the Big Bang. Assume there was a creation event. As the event expanded from a point, the material particles became less dense and began to cool. The primordial material was mostly photons with a very small neutral admixture of protons, neutrons, and electrons. For some reason, matter predominated over antimatter and escaped annihilation, but just barely. The reason for a matter–antimatter asymmetry is not yet understood. The number of photons far exceeds the number of protons, for example.

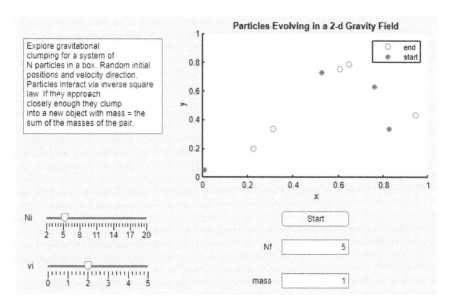

Figure 2.1:   In this particular choice, a low density and a moderate velocity lead to no aggregates at all. Results will vary due to the random initial positions and velocities. Many final configurations are possible.

As long as the weak interactions could remain in thermal equilibrium the mixing of neutrons and protons along with electrons and neutrinos could continue, as seen in the following equation:

$$n \leftrightarrow p + e^- + \bar{v}_e$$
$$v_e + n \leftrightarrow p + e^- \tag{2.1}$$

If thermal equilibrium exists, the number density, n, of protons and neutrons is governed by the Boltzmann factor, which for non-relativistic nucleons is just the proton–neutron mass difference, Eq. (2.2). For high temperatures, $kT \gg Q$, there will be equal numbers of protons and neutrons. At these high energies and temperatures, it is more appropriate to use the electron volt, eV as the energy unit. The mass of an electron is 0.511 MeV, for example. We know $1\,\mathrm{eV}$ is $1.6 \times 10^{-19}\,\mathrm{J}$, then

$$n_n/n_p = e^{-Q/kT}, Q = m_n - m_p = 1.29\,\mathrm{MeV} \tag{2.2}$$

The mass fraction of neutrons, X_n, is the ratio of the neutron number density to the nucleon number density (neutrons and protons are nucleons) as defined in the following equation ignoring the small n-p mass difference compared to the masses of the nucleons:

$$X_n = n_n/(n_n + n_p) = n_n/n_N \qquad (2.3)$$

The time evolution of the neutron mass fraction arises because the proton initiated n creation needs to stay in balance with the neutron decay process. The reaction rate Γ_{np}, for Eq. (2.1), causes contributions to dX_n/dt from protons $(1 - X_n)$ with reduced Boltzmann factor making neutrons and from neutrons X_n losing neutrons. The weak interaction rate is highly temperature dependent, scaling as the fifth power of temperature while the expansion rate of the Universe, the Hubble parameter H, in a radiation-dominated epoch scales as the square of the temperature, Eq. (2.4). The weak interaction strength is specified by the Fermi coupling constant, G_F, where numerically $G_F = 1.6 \times 10^{-5}\,\text{GeV}^{-2}$:

$$dX_n/dt = \Gamma_{np}[(1 - X_n)e^{-Q/kT} - X_n]$$
$$\Gamma_{np} \sim G_F^2 T^5, \quad H \sim T^2 \qquad (2.4)$$

As the expansion and cooling of the Big Bang fireball continued, at some point the reaction rate was not sufficient to keep the protons and neutrons in statistical equilibrium. The decoupling of the neutrons and protons is explored with the "Neutron_Freeze" App. The Hubble "constant" or rather parameter is H and has the dimensions of inverse time.

The relativistic energy density or mass density scales as T^4, as is familiar from the black body energy density of the photons, which also factors into the reaction rate. The photons dominate the particle number density, so the energy density is the relativistic prediction (Stefan–Boltzmann law).

The temperature when neutrons and protons decouple is roughly taken to be the temperature when the reaction rate and the expansion rate are equal, $\Gamma = H$ at T_{fr}. The actual value of this "freeze out" temperature, T_{fr}, is about 1 MeV and is found using the Matlab utility "min" applied to the quantity $\Gamma - H$. The frozen

n/p ratio is equal to $\exp(-Q/kT_{fr})$ after which the protons and neutrons propagate independently.

The reaction rate and the Hubble parameter are plotted in the App "Neutron_Freeze" shown in Figure 2.2. The freeze out temperature and neutron to proton ratio are displayed numerically. This mix of photons, neutrinos, electrons, neutrons, and protons is then set to evolve into heavier nuclei, which will be the medium from which the first generation of stars is formed.

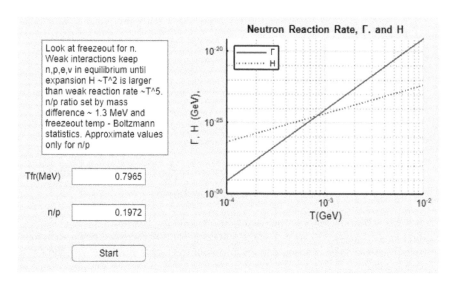

Figure 2.2:   Plot of the reaction rate for $n \longleftrightarrow p$ transitions and the Hubble parameter as a function of temperature, T (kT actually). Electron volts are used as the energy unit at these high temperatures.

2.3. Solar Hydrogen and Helium

For a first-generation star, the composition of nuclei is almost purely protons and helium. Heavier elements are only created in later stellar evolution or in supernova explosions as will be explored later. The binding energy of nuclei is shown in Figure 2.3. Helium has a large binding energy per nucleon (proton or neutron) and is quite stable because it has a closed nuclear "shell" much like the noble gases with closed electron shells. In the primordial formation of nuclei, there is a competition between the Hubble expansion with associated cooling and the binding energy of the produced nuclei which will be essentially stable if the temperature is such that the nucleus in question is not broken up thermally by the dominant photon bath in which the matter is immersed. Helium, with the deep binding is quite stable.

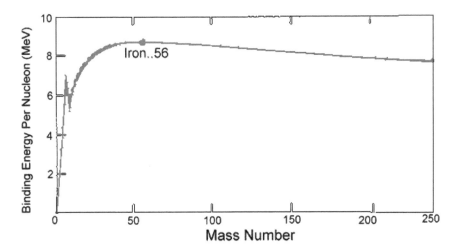

Figure 2.3: Binding energy per nucleon vs. atomic number, A. The largest binding energy occurs for iron. Fusion processes cannot therefore make elements heavier than iron. The binding energy for helium is larger than that for the next heavier nuclei.

The neutron to proton ratio was frozen in when the temperature of the Universe was too small for the weak interactions to keep them in thermal equilibrium. Neutrons are unstable and will decay but on a very long time scale, about 881 sec, compared to the time for helium formation. In fact, they will end up stably bound into the helium nuclei. A chain of fusion processes with intermediate deuterium, D, leading to helium production is given as follows:

$$p + n \rightarrow D + \gamma$$
$$D + D \rightarrow n + H_e^3 \qquad (2.5)$$
$$H_e^3 + D \rightarrow p + H_e^4$$

The protons and neutrons fuse to form deuterium (binding energy, B, 2.2 MeV). Two deuterons fuse to form $He^3 (B = 7.7 \, \text{MeV})$, which then fuses with a deuteron to form $He^4 (B = 28.3 \, \text{MeV})$. The differences in binding energy are crucial because they enter into the exponential Boltzmann factors. Because these fusion processes occur in a dense photon medium, the system entropy is large and the nuclei form at energies, kT, much below the binding energy scales, B.

Define the relative nuclear mass abundance, atomic weight A, by X_A in Eq. (2.6), and recalling the frozen neutron number density with respect to protons, we have

$$X_A = An_A/n_N$$
$$X_n \sim X_p e^{-Q/kT} \tag{2.6}$$

There is a simplified model for the abundance of heavy nuclei which uses only the thermodynamics of the formation process, Nuclear Statistical Equilibrium (NSE), and does not address the dynamic of the fusion processes. The expected helium abundance in this simple model is given in Eq. (2.7), where T is the temperature and η is the fraction of baryons to photons. Clearly, to form a helium nucleus, two protons and two neutrons are needed and the helium binding energy, B_4, plays a crucial role in defining the helium abundance as it appears in the exponent. A large binding energy enhances the helium abundance because once formed it is less likely to be subsequently thermally broken apart, i.e.,

$$X_{\text{He}} \sim T^{9/2} \eta^3 e^{B_4/kT} X_p^2 X_n^2 \tag{2.7}$$

The approximate behavior of the light element abundances of p, n, and He are shown using the App "Light_Elements2" which uses only this simple model. The deuterium abundance is tracked but not plotted because it is tiny due to the small binding energy for D. The system of equations for p, n, D, and He is started off at a kT of 0.8 MeV with the frozen abundances for p and n while the initial D and He abundance is taken to be zero. The equations are then evaluated at lower temperatures. The output of this App is shown in Figure 2.4. The entropy is proportional to B defined in Figure 2.4 to be the fractional nucleon mass to all of the energy in

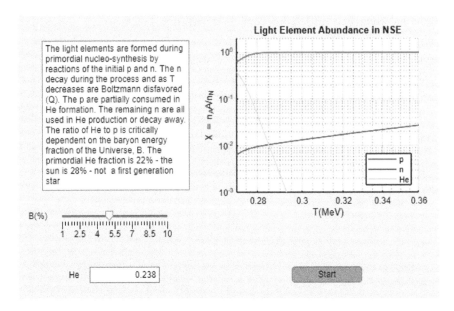

Figure 2.4:   Approximate value of the abundances of p, n, and He as a function of the temperature of the photon medium.

the Universe. Since $B = \Omega_b$, it is proportional to η, dominated by photons, $\eta \sim 1.4 \times 10^{-8}\Omega_b$. Therefore, a measurement of the helium mass fraction gives fundamental information about the composition of the early Universe.

The EditField output called "He" is simply the value for X_4 at 0.271 MeV and not the true final primordial abundance. The fraction of the total energy density of the Universe carried by baryons (nucleons) is a critical parameter driving the fractional helium abundance and is determined to be only about 5%. The Slider can be used to study the dependence of the "He" value on the baryon fraction B in the simple model since B scales the entropy η. The final primordial helium abundance at temperatures where primordial nucleosynthesis effectively ceases is about 22% with 78% p. Other nuclei such as deuterium or lithium are very small so $X_p + X_4 = 1$. All of the neutrons have decayed or are taken up into helium nuclei. The protons are either free or also taken into helium.

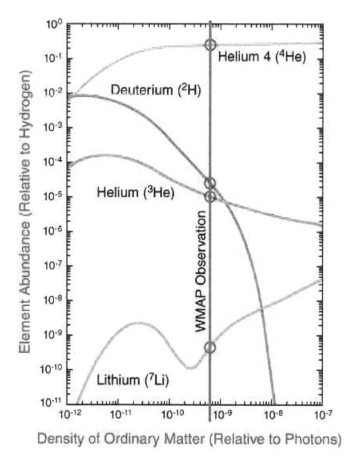

Figure 2.5: Precise calculation of the primordial abundance of nuclei as a function of η, the baryon to photon ratio during nucleosynthesis which can be converted to Ω_b the mass density of the Universe due to baryonic matter.

The results of a much more precise calculation of the primordial light element abundances are shown in Figure 2.5. The only nucleons remaining are dominantly the single protons and the protons and neutrons stably bound into helium. Other nuclei are sub-dominant but supply crucial cross checks for the more correct and detailed calculations used to create Figure 2.5. The helium abundance in turn specifies the baryon fraction of the early Universe. The x axis is $\eta = 1.4 \times 10^{-8}\Omega_b \sim 7 \times 10^{-10}$.

A star is a copious source of neutrinos created in the chain of fusion reactions which occur via electroweak interactions, such as

$$p + p \rightarrow D + e^+ + \nu_e \qquad (2.8)$$

Because the energies of particles in the sun are insufficient to create the other "flavors" of neutrinos, solar neutrinos are a favored place to look for oscillations between the neutrino flavors. A Nobel Prize for the discovery of neutrino oscillations using solar neutrinos was awarded in 2015. However, a good solar model was needed to predict the initial electron neutrino flux correctly when searching for a disappearance of the electron neutrinos during travel to the Earth. For this reason, among others, a section on models of the Sun is included later in the text.

The solar neutrino spectrum for different fusion processes is shown in Figure 2.6. The dominant process is the p–p reaction which yields neutrinos with energies less than about 1 MeV. These neutrinos

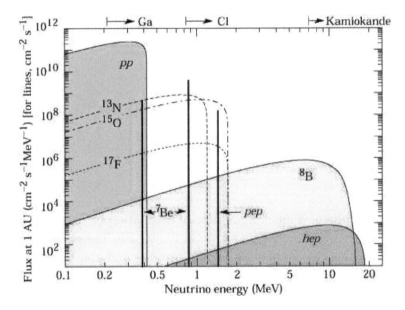

Figure 2.6: Neutrino energy spectrum for neutrinos emitted by the Sun during fusion reactions. The dominant process, by at least three orders of magnitude in rate, is p–p.

can be detected on Earth using Gallium as a target. Higher energy neutrinos from rarer processes can be found using chlorine targets, or for even higher energies using Cherenkov light in water detectors. The Cl to Ar transition has a neutrino threshold energy of 0.8 MeV. The Ga to Ge threshold is 0.23 MeV. Note that aside from the p–p reaction, the indicated sources of high-energy solar neutrinos are all due to the appearance of elements in the Sun which are not primordially produced.

2.4. Protostars — Star Formation

At the end of primordial nucleosynthesis, the Big Bang plasma of photons, protons, and electrons has become a mixture of neutral hydrogen and about 20% helium with only traces of a few other light elements. This early interstellar medium (ISM) is the cradle for the formation of the first generation of stars. The neutrons are all captured and stabilized in the helium nuclei or have decayed. The existences of small anisotropies in the cosmic microwave background (CMB) are the seeds of large-scale structure and their observation led to the award of the Nobel Prize in 2010.

The ISM is largely neutral hydrogen, H. A cloud of hydrogen will vary in density from place to place ultimately due to fluctuations in the isotropy of the CMB. Gravity has a mean self-energy $\langle V \rangle \sim -GM^2/R$ for a part of the cloud with a mass M contained in a radius R. The cloud has a mean thermal kinetic energy $\langle K \rangle \sim NkT$, where N is the number of protons $\sim M/m_p$. A region of the cloud collapses if $2\langle |V| \rangle / \langle K \rangle$ is approximately greater than 1 and gravity dominates. This factor will appear again later in the discussion of orbits for bound Keplerian systems. Dropping all factors of order 1 in order to get an order of magnitude estimate, $GM_J^2/R \sim \langle K \rangle = NkT = M_J kT/m_p$, where M_J is the Jeans mass, the largest mass that is unstable to gravitational collapse. Taking $M_J \sim \rho_o R^3$, the Jeans mass becomes $M_J \sim (kT/Gm_p)^{3/2}/\rho_o^{1/2}$, which depends only on the temperature and the initial density. For example, at a temperature of 10 K and a Jeans mass of one solar mass, the critical density would be $\sim 5 \times 10^{-16}$ kg/m^3. As expected from Section 2.1, low densities or high temperatures lead to higher Jeans masses.

Another way to consider the criterion for gravitational collapse is to consider when fluctuations can be stabilized by pressure waves since such waves respond at the speed of sound in the medium. For a fluctuation over a large enough volume, the sound waves cannot cover the entire region. The isothermal speed of sound is approximately given by c_s in Eq. (2.9). The region collapses if the mass within it, of density ρ_o, is greater than the Jeans mass, M_J. The wave number of the sound wave is k_λ. The dimension of wave number times velocity is inverse time (ω). The characteristic time for a gravitational response is the square root of $G\rho_o$. The Jeans length is that length where the two length scales are equal. These two approaches agree within a numerical factor of order 1:

$$(k_\lambda c_s)^2 - 4\pi G\rho_o > 0, \quad c_s = \sqrt{P_o/\rho_o} = \sqrt{kT_o/m_p}$$
$$\lambda_J = c_s\sqrt{\pi/G\rho_o}, M_J = \rho_o(\lambda_J/2)^3 = \rho_o[(\pi kT_o)/(4m_pG\rho_o)]^{3/2}$$

$$(2.9)$$

The Jeans mass scales as the inverse square root of the density, and the 3/2 power of temperature so that hot and diffuse regions are stable against collapse. Denser regions are unstable and will serve as the seeds of stellar formation. The scaling in T and density of the Jeans mass was also "derived" using the above-mentioned energy arguments. These results can be considered to be the quantitative formulation of the results of Section 2.1, where low density clearly inhibits "gravitational clumping".

The collapsing region is in free fall with acceleration of $-GM(r)/r^2$. Integrating the force equation with zero initial velocity, one can use energy conservation to find dr/dt and then integrate again to find the time to collapse to a point in Eq. (2.10). That time depends only on the initial density. Dimensional analysis, or λ_J/c_s, also implies that the collapse time should scale as the square root of $1/(G\rho)$. For example, with an initial density of $3 \times 10^{-17}\text{kg/m}^3$, the time is about 400,000 years:

$$dr/dt = \sqrt{2GM(1/r - 1/r_o)}, \quad \tau_c = \sqrt{3\pi/32G\rho_o} \qquad (2.10)$$

For a pressure-less sphere of density ρ_o the time to collapse, for example, for sphere of density 1 kg/m^3, is about 18 h.

The collapse of a mass near the mass of the Sun is explored in the App, "Protostar_Formation". An initial region at a temperature of 10 K made up only of protons with a number density of 5×10^{10}/m^3, mass density 8.35×10^{-17} kg/m^3, is defined. The collapse may or may not begin based on the mass Slider value chosen by the user. The initial mass value is set by the Slider and the initial mass and radius are compared to approximate values for the Jeans mass and radius. If the Jeans radius is sufficient for stability, no collapse is reported. There is no collapse for masses greater than about 2.2 solar masses in this example.

If collapse begins, it is tracked to increasing temperature and density to the point where the hydrogen is ionized releasing the binding energy, approximately 13.6 eV. Further increases in temperature occur until the electron Fermi pressure stabilizes the star against further collapse. The discussion of Fermi pressure is deferred for now until it is taken up in the later discussion of the collapse of stars into "white dwarfs". Suffice it to say for now that the Fermi/Pauli exclusion principle does not allow the quantum wave functions of the electrons, of spatial extent comparable to the de Broglie wavelength, to overlap and that creates a pressure which halts the gravitational collapse. The wave associated with a particle has a wavelength $\lambda = h/p$ and a frequency $v = \varepsilon/h$, where h is the Planck constant, p is the particle momentum, and ε is the energy.

The figure from the App "Protostar_Formation" is shown in Figure 2.7. The Text box shows a summary of the App setup. The Slider sets the initial mass, and the initial radius is compared to the Jeans radius. For the values shown here, there is a collapse. The collapse to a point takes about a quarter of a million years. However, before that the electron Fermi pressure halts the collapse and stabilizes the protostar at a radius R_F with a temperature T_F. Indeed, the final temperature is not so far away from that found at the core of the Sun. This implies that fusion can begin and the star may evolve to a steady state. Lower Slider masses lead to lower final temperatures.

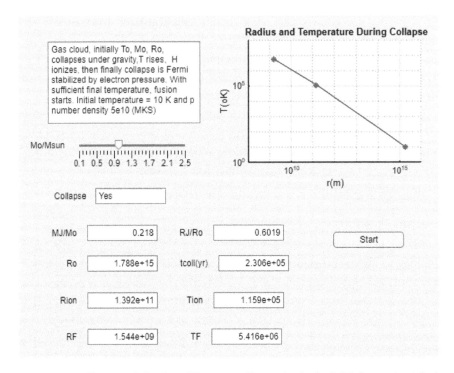

Figure 2.7:   Output of the App "Protostar_Formation". An initial mass is picked and the collapse is followed in time as the radius decreases and the temperature rises.

But first, the fusion process must get started. The initial fuel is only protons since the deeply bound helium is inert. Production of deuterium is a slow and weak process. The initial reactions resulting in helium and two protons, Eq. (2.11), keep the cycle going:

$$p + p \rightarrow D + e^+ + \nu_e$$
$$D + p \rightarrow H_e^3 + \gamma \qquad (2.11)$$
$$H_e^3 + H_e^3 \rightarrow H_e^4 + p + p$$

However, the protons repel one another, and to overcome this "Coulomb barrier" classically a very high temperature is needed to achieve a separation of about 10^{-15} m (1 Fermi), which is the typical size of a nucleus, where the nuclear fusion process can be initiated. An estimate for the necessary classical temperature T_c is that $(3kT_c)/2 = (e^2/r)/(4\pi\varepsilon_o)$ or $T_c \sim 10^{10}$ K, which is not available. How does the protostar start to begin fusion? It is by quantum mechanical tunneling, Eq. (2.12), of the protons through the Coulomb barrier. The de Broglie wavelength, $\lambda = h/p$, relates the quantum particle momentum and the size of the wavefunction. The required quantum temperature is estimated by equating the proton kinetic energy and the Coulomb repulsive energy as follows:

$$\lambda = (2\pi\hbar)/p$$

$$e^2/(4\pi\varepsilon_o\lambda) = p^2/(2m_p) \tag{2.12}$$

$$kT_{QM} = (4e^4 m_p)/[3(8\pi^2\varepsilon_o\hbar)^2]$$

Numerically, T_{QM} is about 10^7 K, which is the right order of magnitude. The stars start to shine due to quantum tunneling. They would not be thought to be able to begin fusion if considered to be purely classical objects.

A representation of the numerical results for energy production in the p–p reactions depends on the density, the mass fraction of the protons, X, and the temperature — to a high power of temperature, since the Coulomb barrier still exists even with tunneling. An approximate representation of the energy generation parameter, ε_{pp} in power per unit mass, near the temperature of the core of the Sun for p–p reactions is given as follows:

$$\varepsilon_{pp} \sim (1.1 \times 10^{-12})\rho X^2 (T/10^6)^4 (W/kg) \tag{2.13}$$

2.5. Constant Density "Star"

The simplest star model is a uniformly dense body in thermal equilibrium. The simplicity is a good introduction but is not a reasonable representation of the details of the interior stellar structure. The composition of the star, assumed to be first generation, is primordial, mostly protons with ~20% helium. The fusion reaction which was started at the end of the Jeans collapse provides radiation pressure which is in thermal equilibrium with the pressure of gravitation. The equations that describe the star at this stage of its life cycle, ignoring the helium for now, are as follows:

$$M = 4\pi\rho R^3/3, \quad P(r) = 2\pi G\rho^2(R^2 - r^2)/3$$
$$PV = NkT, \quad \rho = Nm_p/V \tag{2.14}$$
$$P/\rho c^2 = kT/m_p c^2$$

With a constant density, ρ, the total mass is just the density times the spherical volume. The pressure as a function of radius follows from the relationship between pressure, density, and mass, $dP/dr = -GM(r)\rho(r)/r^2$ for a constant ρ so that $M(r) = 4\pi r^3 \rho/3$. An ideal gas composed of only protons is assumed, with number N, volume V, and number density $n = N/V$. The mass density is just $\rho = nm_p$. The quantity $P/\rho c^2$ is in this simple case only a function of temperature, T, and is dimensionless.

Stars are stable over a wide range of masses and densities. The App used to look at this simple star model is called "Star_Const_Rho". The Figure made by the App is shown in Figure 2.8. A Slider is chosen to pick the stellar density. The core pressure, at r of zero, is computed and plotted as a function of stellar mass as is the radius. Both pressure and radius are scaled to that of the Sun as is the mass. The ratio $P/\rho c^2$ is small for typical stars, but since in general relativity, GR, all energy gravitates, it will be shown later that the ratio is not small for degenerate stars. The display of $P(r)$ is not accomplished using symbolic variables and "char" as in previous examples, but simply by writing a text string and displaying it in the EditField as an alternative method. The plots shown in Figure 2.8 show that the constant density model of the Sun is a good first approximation.

In general, as a rough guide the scaling of stars follows from the approximate results on Eq. (2.14) and are shown in Eq. (2.15). The density scales as M/R^3. The pressure scales as $(\rho R)^2$ or $(M/R^2)^2$. The temperature scales as PV or M/R. Finally, the luminosity scales roughly as the cube of the mass, which is motivated in what follows when it will be seen that the fusion processes driving the luminosity are very temperature dependent and are localized close to the core

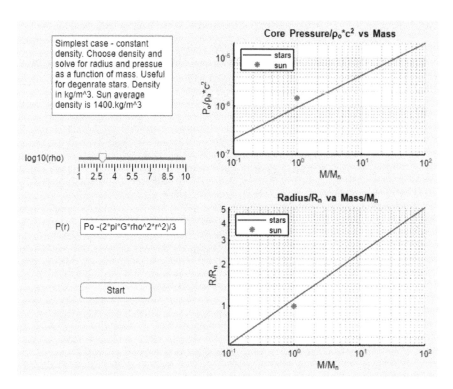

Figure 2.8: Figure from the App "Star_Const_Rho". A density is set by the slider, which defines the radius for a given mass and the core pressure. Values for the Sun are indicated, showing that the Sun is a typical star.

of the star where the temperature is highest:

$$\rho \sim M/R^3, \quad P \sim M^2/R^4, \quad T \sim M/R, \quad L \sim M^4 \tag{2.15}$$

In fact, the observed dependence is closer to the fourth power of mass, as is seen in Figure 2.9, where data on some main sequence stars are shown.

Parenthetically, Matlab uses log to evaluate the natural logarithm, ln, and log10 for base 10. To represent the exponents, aeb in Matlab is $a \times 10^b$, while a.* $\exp(b)$ represents ae^b. These conventions in Matlab may take a while to get used to.

Main sequence stars have a relationship between luminosity, total power, and mass. They are about 90% of all known stars. They burn protons to make helium and have masses from about 0.1 to 100

solar masses. The luminosity of a main sequence star scales, very
approximately, as M^4. A plot of some main sequence stars is shown
in Figure 2.9. The log–log plot illustrates the approximate power law
dependence of luminosity, L, on mass, M.

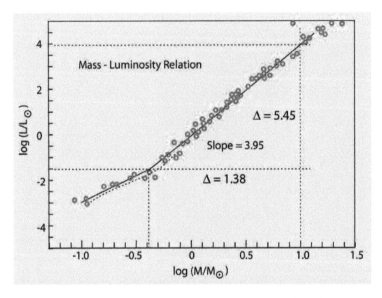

Figure 2.9:   Plot of the mass of main sequence stars vs. their luminosity with
both M and L scaled to the values of the Sun.

2.6. Solar Temperature and Color

Main sequence stars have a variety of colors: blue, yellow, and red
are common. There is a connection between the color of a star and
the temperature of a star, which is defined as that which for a black
body gives the luminosity of the star. This is the effective surface
temperature of the star, T_{eff}.

For a body at a given temperature, T, there is a distribution of
particle energies, ε, which can be measured. The basic idea is that all
momentum components are equally probable. The joint probability
to find p_x, p_y, and p_z is then proportional to the product of the
probabilities $dp_x dp_y dp_z$. If there is no preferred direction for the
system, this product is the probability to find momentum p between

p and $p + dp$, which is just $2\pi p^2 dp$. Since $\varepsilon = p^2/2m$ for a non-relativistic (NR) material particle, one can change variables to find the probability to find an energy ε between ε and $\varepsilon + d\varepsilon$, using a change of variables, $d\varepsilon = pdp/m$. The result is $2\pi(2m\varepsilon)(md\varepsilon/p)$, which is $2\pi\sqrt{(2m\varepsilon)}d\varepsilon \sim \sqrt{(\varepsilon)}d\varepsilon$.

The Maxwell–Boltzmann probability for the energy, ε, to be observed at a given temperature, T, is then given a weight $\exp(-\varepsilon/kT)$ because at that temperature all momentum components are not equally probable. A constant is also determined to normalize the probability to 1, which just means you will always find some particle energy, ε such that the total number density is n. The units of Eq. (2.16) are inverse energy times volume, i.e.,

$$d(n(\varepsilon))/d\varepsilon = [(2n\sqrt{\varepsilon/\pi})e^{-\varepsilon/kT}]/(kT)^{3/2} \qquad (2.16)$$

The distribution of the energy density, u, for photons follows the Bose–Einstein statistics as a function of photon energy ε and temperature, T, has dimensions of inverse volume, and is given as

$$du/d\varepsilon = [(kT)^3/\pi^2(\hbar c)^3][x^3/(e^x - 1)], \quad x = \varepsilon/kT \qquad (2.17)$$

The Maxell–Boltzmann weight no longer applies because here the energy density refers to photons and they obey the statistics of even–spin quantum particles, the Bose–Einstein statistics. If x is much greater than 1, then the factor $1/(e^x - 1)$ approaches the e^{-x}-factor of classical Maxwell–Boltzmann statistics appropriate for situations where the particle energy is much greater than the thermal energy $\sim kT$.

Integrating over all photon energies, ε, at fixed temperature, the energy density, $u(T)$, and the flux, $f(T)$, are given as

$$u(T) = (4/c)\sigma T^4, \quad \sigma = \pi^2 k^4/(60\hbar^3 c^2)$$
$$f(T) = \sigma T^4 = L/A, \quad P = 4\sigma T^4/(3c) = u(T)/3 \qquad (2.18)$$

The energy density, light pressure, P, and flux, luminosity per unit area, all scale as the fourth power of the temperature and the factor for the flux or power crossing unit area, σ, is called the Stefan–Boltzmann constant. The pressure due to the photons is directly proportional to the energy density, $P = u/3$.

The distribution of photon wavelengths at a fixed temperature can be derived from Eq. (2.17) by a change of variables from energy to wavelength, with frequency f, circular frequency ω, and wavelength λ using the relationship $\varepsilon = (hc\lambda)$ where h is Planck's constant:

$$\varepsilon = \hbar\omega, \quad \lambda f = \lambda(\omega/2\pi) = c$$
$$du/d\lambda = 16\pi^2(\hbar c)/[\lambda^5(e^{2\pi\hbar c/\lambda kT} - 1)] \tag{2.19}$$

For the Sun, the surface temperature, T_{eff}, is about $5,780°$. The solar flux, f, in W/m^2 is about 6×10^7 and the peak wavelength is at about 500 nm. This flux as intercepted at the orbit of the Earth is about $1\,\text{kW/m}^2$ and is called the solar "constant". The flux actually striking the solar cells on the Earth's surface after having traversed the atmosphere is somewhat less.

The connection of temperature and wavelength for a star is examined in the App "Star_T_Lam". A Slider is used to set the temperature. The distribution $du/d\varepsilon$ is treated symbolically using the "syms" declaration. It is integrated to obtain u using "int" and displayed using "char" in an EditField. The wavelength distribution is also displayed symbolically using "char". It is then evaluated numerically using the utility "eval" and a plot of the spectrum is made. The maximum of the spectrum is found using "max". The numerical values of the solar flux and the solar constant are displayed as well as the wavelength of maximum emission. Using the Slider, the user can vary the blackbody "color" of light over a wide range of peak wavelengths from about 900 nm at low T to about 200 nm at high T Slider values. The initial display Figure is shown in Figure 2.10. As usual, a brief explanation to make the App somewhat stand-alone of the App is provided in the Text box.

2.7. Star ODE

In order to make a better approximation to the interior of a star than simple uniformity, the differential equations approximating the thermal equilibrium for the star need to be used. A schematic diagram of the solar interior is shown in Figure 2.11. The core of the Sun is at the highest temperature and fusion largely proceeds there. High temperatures are needed because many fusion reactions

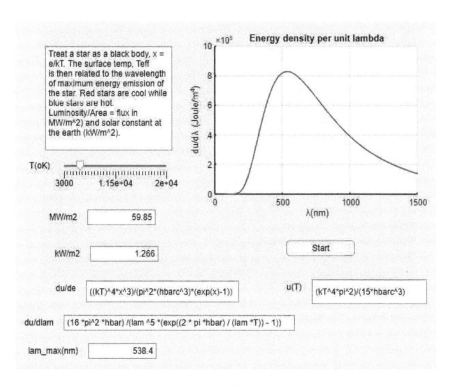

Figure 2.10: Output figure for the App "Star_T _Lam". The user picks a surface temperature and the solar flux and solar constant are displayed as well as the wavelength at maximum emission. The symbolic forms of the energy density $du/d\varepsilon$, $du/d\lambda$, and $u(T)$ are displayed using EditFields.

involve charged particles that Coulomb repel one another and high velocities are needed to overcome these Coulomb barriers. Even aided by quantum tunneling, temperatures of a few million degrees are needed.

Outside the core, the radiation pressure of the photons is the dominant force that balances the gravitational attraction of the solar materials. For the Sun, this region is the largest part of the interior. Outside the radiative zone, convective motion is the dominant physical effect. Outside of the convective zone, the "surface" of the Sun has a complex behavior. For example, the rotating plasma of charged particles in the solar interior creates a magnetic field which is evident in the behavior of solar prominences. These details are of interest but are a bit too detailed to be covered here.

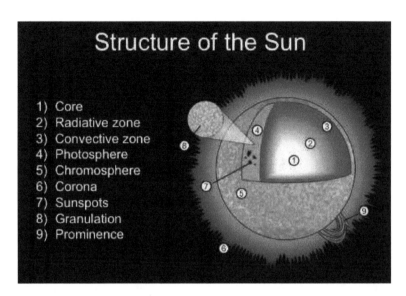

Figure 2.11: Schematic diagram of the interior structure of the Sun. The Sun is assumed to be in thermal equilibrium where the gravitational attraction is balanced by radiation pressure.

To begin, the Sun is treated as an ideal gas with the majority constituents of mass being the protons and helium ions since the hydrogen and helium are ionized at core temperatures corresponding to MeV energies compared to eV ionization energies. The core is then a plasma of NR protons, electrons, and helium nuclei. The energy density is the product of the number density times the mean thermal energy, $(3kT)/2$, i.e.,

$$PV = NkT = nRT, N/V = n$$
$$P = (n/3)m\langle v^2 \rangle \qquad (2.20)$$
$$m\langle v^2 \rangle/2 = 3kT/2$$

In general, there are contributions to the pressure from the gas and from the radiation. The mass density is related to the number density by μm_p, where μ is the mean molecular weight. The mean molecular weight is the average mass of a particle in the gas, so that $\mu m_p \sim 0.62$ for the mix of protons and helium nuclei of a young star when fully ionized, $1/\mu \sim 2X + 3Y/4$, where X is the proton mass fraction and Y is the helium mass fraction. The mean mass is less than a proton mass because of the existence of three electrons in the ionized mixture. The neutral fraction when the gas is not ionized is $1/\mu \sim X + Y/4 = 1/1.3$ larger than the proton mass due to the heavier helium with A of 4 and the lack of free electrons:

$$P = \rho kT/(\mu m_p) + 4\sigma T^4/(3c) \tag{2.21}$$

The simplest solar model assumes a steady-state stable solution for the star, which is characterized by a density ρ, a pressure P, and a temperature T, all functions of the radius r from the solar center. The star is in equilibrium between the gravitational pressure and the pressure due to the gas and the radiation. In the notation used here, $M(r)$, $\rho(r)$, $P(r)$, and $T(r)$, is assumed where M, ρ, P, and T are used for simplicity to denote the mass within r and the density, pressure, and temperature at r. Typically, the pressure is dominated by the contribution due to the gas and not the radiation.

The core where fusion occurs is only about 10% of the mass of the Sun. In fusing protons into helium, only about 0.007 of the rest mass is converted to energy. Typical binding energies are 7 MeV, as shown in Figure 2.3, while typical light nuclear masses are 1,000 MeV, ratio = 0.007. An estimate of the total possible energy release is then $0.1 \times 0.007 \times M_o c^2$ or about 1.3×10^{44} J. Compared to the solar luminosity, 3.8×10^{26} W, a crude estimate of the solar lifetime is about 10 billion years. So, the Sun is expected to burn stably for a good fraction of the time the Universe has been in existence, currently estimated to be about 13 billion years. In fact, until fusion processes were understood, physicists were unable to say how the Sun could have been stable over the long time spans required by the geologists and evolutionists.

The basic equations for the simple solar model, ignoring the radiation pressure for now, are as follows:

$$dM(r)/dr = 4\pi r^2 \rho$$
$$dP/dr = -GM(r)\rho/r^2$$
$$P = (\rho kT/\mu m_p)$$
$$M(0) = 0, M(R) = M_T, P(R) = T(R) = \rho(R) = 0$$

$$\tag{2.22}$$

In the equations, $M(r)$ is the total mass inside radius r, while T, P, and ρ are the values of temperature, pressure, and density at r. The radiation pressure term is $(4\sigma T^4)/3c$, which could be added to the pressure, but the contribution to the pressure is small in most cases. The equation for the mass r dependence on density is clear, while the dependence of pressure on G, M, and density was mentioned previously, Eq. (2.14), $dP/dr \sim -G\rho^2 r$, $\rho \sim M/r^3$. The temperature dependence follows from the ideal gas law for pure protons and is altered in Eq. (2.21) for a mixture of ions. The boundary conditions at the core and the surface define the meaning here of solar radius, R. The quantity μ is the mean molecular weight, $\mu = \langle m \rangle / m_p$. Strictly speaking, the mass should be that of hydrogen but that is so close to the proton mass as to be just numerically insignificant. The value of μ depends on the gas described and the state of ionization. For the Sun with H and He mass fractions X and Y, the quantity $1/\mu$ is about 0.62.

The specific heat is defined to be the amount of heat needed to raise the temperature of an object by $1°$. If the processes are adiabatic, then there is no net heat flow. Rising gas expands and cools. Falling gas contracts and heats. An adiabatic process has PV^γ constant. A convective temperature behavior with $\gamma = 5/3$ relates T and P as a function of r. The factor γ is the ratio of $-dP/P$ divided by dV/V, which is the ratio of the specific heats and has a value $5/3$ for an ideal gas. Both the case of simple convective dominance and simple radiative dominance are explored with the present App but nothing more complex is attempted.

In the case of radiative dominance, the photons scatter off the charged protons, electrons, and H ions. The scattering is characterized by an opacity, κ, and a luminosity, $L(r)$. For a density ρ and path length z, the photon intensity falls exponentially with path length z with a dimensionless exponent of $-\kappa\rho z$. The mean free path is $n\sigma = \kappa\rho$, where σ is the cross section for light scattering off the solar medium. The two options for dT/dr are shown in the following equation:

$$dT/dr|_{\text{conv}} = [(\gamma - 1)/\gamma](T/P)dP/dr$$
$$= (1/\gamma - 1)(\mu m_p GM(r))/(kr^2) \qquad (2.23)$$
$$dT/dr|_{\text{rad}} = -(3\langle\kappa\rangle\rho)(L(r)/4\pi r^2)/(16\sigma T^3)$$

The luminosity behavior, $dL(r)/dr$ follows that of the mass, but is modified by an energy production factor, ε_{fus}, which, since it represents fusion, is highly temperature dependent and thus biased toward the inner, high-temperature, core. In this App, the opacity, with dimension m^2/kg, is set to a constant times ρ divided by $T^{-3.5}$. The fusion energy production factor is taken to be that for p–p reactions, Eq. (2.13). For the Sun, the luminosity to mass ratio implies a mean power production of about 0.2 mW/kg. At the core, using Eq. (2.13) the power production is the highest, about 4.3 mW/kg due to a core number density of protons of about $9 \times 10^{31}/m^3$:

$$dL(r)/dr = 4\pi r^2 \rho(r)\varepsilon = [dM(r)/dr]\varepsilon_{\text{fus}} \qquad (2.24)$$

These equations are solved in the App script "Sun_ODE" in either the purely convective or purely radiative case. The solution consists in choosing the numerical values for two core variables, pressure and temperature, found using a real solar model in the literature and then integrating over increasing radius. Density is derived using Eq. (2.22) and the mass follows using the density. The initial conditions at r of zero determine the solution. Those at R cannot then be imposed.

For the Sun, the fusion occurs in about the first inner ~30% of the radius, where radiation pressure dominates. Convection is important in the outer layers. The figures generated by the App in the convective case are shown in Figures 2.12 and 2.13. The agreement with a full solar model is not too bad. Neither are the plots if the radiative case is chosen. Neither of the sets of plots is particularly promising. Indeed, the core parameters have been assumed, which means the model does not arise from first principles. The radial

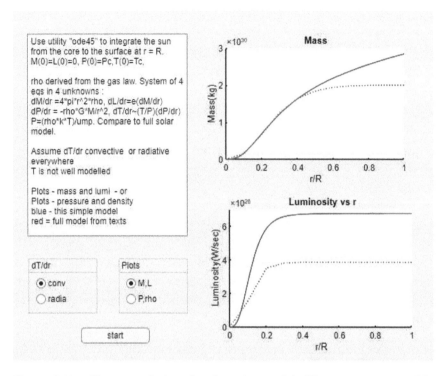

Figure 2.12: First set of plots for the solar model. The mass is reasonably modeled in the core but does not fall off sufficiently rapidly with r. The luminosity is more central than the mass distribution, as expected.

behavior at $r = R$ is not imposed, so the falloff at the solar surface is not automatic.

Two alternatives will now be examined. In the first case, an additional constraint on the solar behavior is imposed, which defines a radial shape. In the second case, the problem is, more correctly, treated as a boundary value problem and the boundary values shown in Eq. (2.22) are imposed. Both techniques give improved agreement with the more correct solar model that was taken from the literature.

2.8. Polytropic Star Model

Integrating out from the core of the Sun is not very satisfying, since an initial core temperature and pressure or density must be assumed

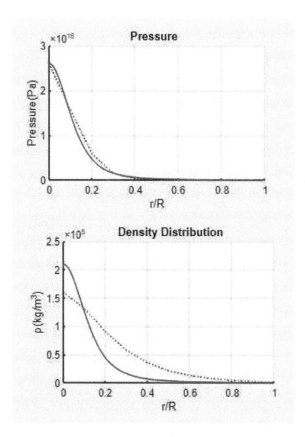

Figure 2.13: The second set of plots generated by the App "Sun_ODE". The pressure falloff is well modeled. The density is derived assuming the ideal gas law. The poor agreement reflects the poor modeling of the temperature distribution.

to be known. Another approach is to assume an additional power law relationship between the variables that define the solar equation of state.

Scaled dimensionless variables for mass, radius, pressure, and temperature are first defined as m, x, p, and t, Eq. (2.25). The notation now distinguishes between the total mass, M_T, and the mass inside r, $M(r)$ as is also true for luminosity, $L(r)$. The quantities m, x, p, and t are all dimensionless as is conventional for polytropes and useful in general, for example, in a BVP treatment of the Sun. The exact coefficients for p and t in Eq. (2.25) are just chosen for simplicity:

$$M(r) = mM_T$$
$$r = xR$$
$$P = p[GM_T^2/4\pi R^4] \tag{2.25}$$
$$T = t(\mu m_p/k)[GM_T/R]$$

Scaled equations and boundary conditions defining the star can be found after a bit of algebra. The conditions are placed on the mass at the origin and the surface at $r = R$ and the vanishing of the surface pressure and temperature. Only the equation for convective temperature dependence is shown here. Note that dp/dx has a singularity at $x = 0$, which needs to be evaded. All constants have been absorbed in the definitions of the dimensionless variables:

$$dm/dx = (p/t)x^2$$
$$dp/dx = -pm/(tx^2)$$
$$dt/dx = (2/5)(t/p)(dp/dx) = -(2/5)(m/x^2) \tag{2.26}$$
$$m(0) = 0, m(1) = 1, p(1) = 0, t(1) = 0$$

Better models can then be explored which will make improved approximate starting values to the real complex models found in textbooks. For a star like the Sun, dominated by radiative pressure near the core and not convection, the equations are modified as shown in Eq. (2.23). The relationship for dM/dr and dP/dr was already quoted in Eq. (2.22). The expression for dT/dr appears in Eq. (2.23) and that for dL/dr in Eq. (2.24). A scaled quantity for luminosity is also defined using L_o as follows:

$$dL(r) = \varepsilon(\rho, T)dM(r)$$
$$L(0) = 0 \tag{2.27}$$
$$L = lL_o$$

In general, a polytrope relates pressure and density as, $P \sim \rho^\gamma$. For an ideal gas, $\gamma = 5/3$. It is conventional to define an equation of state where the pressure is assumed to be proportional to a power $(1 + 1/n)$ of the density and to depend only on the density, where

$n = 3/2$ for an ideal gas. For the popular choice $n = 3$, the power is $\gamma = 4/3$. The polytropic index is defined to be n. The density is scaled from a core value by the dimensionless function θ, and therefore, the pressure is also scaled to a core value with a power of the index as shown in Eq. (2.28). The core temperature is a derived quantity following from the ideal gas law, i.e.,

$$p = K\rho^{(1+1/n)}$$

$$\rho = \rho_c\theta, \quad p = p_c\theta^{(1+n)}, \quad r_n^2 = (n+1)p_c/(4\pi G\rho_c^2) \qquad (2.28)$$

$$T_c = (\mu m_p)(p_c/k\rho_c)$$

The function θ is a maximum at the origin and falls smoothly to zero at x_R, which defines the solar surface. It is found by numerically solving a second-order differential equation, using the Matlab utility "ode45". The radial shape of θ is the shape of the density. The shape of the pressure is the shape of a power of θ:

$$r = r_n x, \quad \theta(0) = 1, \quad d\theta/dx|_{x=0} = 0$$

$$d/dx[x^2(d\theta/dx)] = -x^2\theta^n \qquad (2.29)$$

$$\theta(x_R) = 0, \quad R = rx_R$$

The polytropic model of the Sun is solved in the App "Polytrop_Sun_App". The Sun is specified by the mass, M_T, and radius, R, and by the index that is chosen by using the Slider. The Sun is well described with an index of about 3, which the user can verify by looking at the results as a function of n, i.e.,

$$M_T = 4\pi R^3 \rho_c[-1/x(d\theta/dx)]_{x_R}$$

$$M(r)/M_T = [-x^2(d\theta/dx)]/[-x^2(d\theta/dx)]_{x_R}$$

$$r/R = x/x_R \qquad (2.30)$$

$$p_c = 1/[4\pi(1+n)][GM_T^2/R^4](d\theta/dx)_{x_R}^2$$

The first set of figures from the polytropic App is shown in Figure 2.14. The user can change the index from the default value of 3 using a Slider. The radial plot of θ for a few indices is seen in the upper left plot. The EditFields give the resulting core pressure,

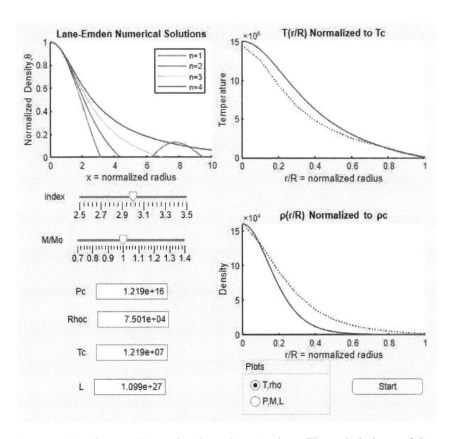

Figure 2.14: Output Figure for the polytropic App. The radial shape of θ is shown, the core numerical values are shown, and the radial distributions of temperature and density are plotted. The blue line is the calculation, while the red dashed line comes from a more correct model.

density, and temperature along with the luminosity for the Slider-chosen value of solar mass and the fixed value of the solar radius. The initial two plots are for temperature and density as a function of radius. In the case of the plots, in the interest of emphasizing the radial shape, the normalization is made to the values of the Sun. However, the values displayed in the EditFields are those arrived at in the polytropic model.

A second set of three plots are available setting the Plots Radio-Group, and the resulting plots for pressure, mass, and luminosity are reproduced in Figure 2.15. It is clear that the luminosity peaks at smaller values of r than the peak for the mass because the energy generation factor in $L(r)$ is strongly temperature dependent. The effect is to make the luminosity more central than mass itself. The shapes of mass, density, pressure, and temperature are in better agreement in this polytropic treatments than in the simple ode treatment shown earlier.

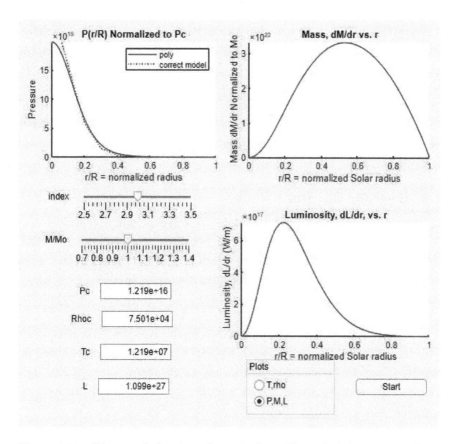

Figure 2.15: Plots made by the polytropic App. The radial dependence of the pressure, mass, and luminosity is displayed. The solar energy production occurs deep in the core of the Sun.

In this App, the Matlab utility used for the system of differential equations is "ode45" as was used in the App "Sun_ODE". For the luminosity, the same function of density and temperature, Eq. (2.13), is used as was used previously in the "Sun_ODE" App. The results, although the machinery is more complex, are a clear improvement over simply assuming core values and integrating out to an ill-defined surface. The polytropic radial shape and the sharp surface work rather well. Indeed, polytropic models are in use by astrophysicists because of their relative simplicity, if only as starting values for more sophisticated models.

An interesting aspect of the App is that M can be varied slightly away from the value of the Sun and the resulting luminosity can be examined. A strong dependence of L on M is observed. Typically, L scales as approximately the fifth power of M for small excursions in mass at fixed radius. This behavior is in reasonable agreement with the behavior of main sequence stars, Figure 2.9, where the observed power is more like the fourth power.

2.9. BVP for the Sun

It remains a fact that the problem to solve is not a classical set of differential equations with initial conditions. It is better suited to a treatment of the Sun as a boundary value problem. The polytropic approach partially alleviates the issue by defining a solar radius. A simple introductory example of a BVP solution was shown previously in Section 1.4 and in Figure 1.14.

The equations for the dimensionless mass, pressure, and temperature in terms of differentials with respect to the dimensionless radius are the same as those used in the polytropic case, Eqs. (2.25) and (2.26). The Matlab utility used now is "bvp4c" with initialization "bvpini" applied to the Sun in "Solar_BVP_App".

The code treats the star this time as a purely radiative object. The equation for dT/dr has been shown already for the purely convective and radiative cases, Eq. (2.23). The opacity as a function of density and temperature and the energy production are needed for the dT/dr equation in the purely radiative approximation and are the same as those used in the App "Sun_ODE". The scale factors for pressure P_s, luminosity L_o, and temperature T_s are the coefficients already shown in the section of polytropic models, defining dimensionless variables, Eqs. (2.25) and (2.27):

$$\kappa = 3.7x10^{18} \rho/T^{3.5} (\text{W/kg})$$
$$dt/dx = 3\kappa(pl/x^2t^4)[\Gamma_s \mu m_p L_o]/[64\pi\sigma RkT_s^5]$$

$$(2.31)$$

These boundary conditions are over-constrained, $m(0) = 0$, $m(1) = 1$, $p(1) = t(1) = \rho(1) = 0$, so a choice is made to release the condition on the surface temperature. However, this assumption makes the problem very sensitive to temperature variations near zero values. The temperature dependence is very strong, leading to some instability in the numerical integrations. With sufficient effort, this issue could be solved. Perhaps, using the polytropic results as starting values would help. It is a detailed effort which will not be pursued here.

As with most boundary value problems, the initial guess as to the solution is quite important. The output figure for the App is shown in Figures 2.16 and 2.17, which show plots of the several variables. In general, the shapes as a function of radius are in improved agreement with the more accurate solar model. Due to the radiative assumption, the BVP variables are m, p, t, and l, while rho is a derived variable using the ideal gas law.

In Figure 2.16, the EditFields display the numerical BVP solutions normalized to the accepted values for the Sun. The shape agreement is not too bad. For the plots displayed in Figures 2.16 and 2.17, the scaled variables are plotted in order to explore the shape rather than the overall normalization. The mass shape is acceptable. The temperature is not required to vanish at the solar surface and has a region that is poorly behaved. The luminosity shape is sufficient and is somewhat more centrally located than was seen in the polytropic case. In Figure 2.17, the pressure and density shapes look quite good. The shape for dL/dr peaks at a value of r/R of about 0.1 which is more central than in the polytropic solution where convective dominance was assumed. There is also a region of instability that reflects the bad behavior of the temperature itself.

For the Sun, radiation dominates for the inner 0.7 of R and the core fusion occurs for radius less than about 0.3 R. The luminosity

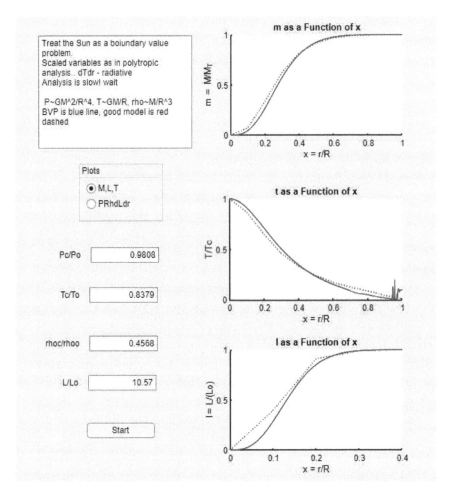

Figure 2.16:   Numerical results for the scaled pressure, temperature, density, and luminosity and plots for the radial dependence of the mass, temperature, and luminosity. Note that $T(R)$ is not forced to be 0, which leads to some instability in the results.

peaks at a radius $\sim 0.1\ R$ and returns to near zero at a radius $\sim 0.2\ R$. This behavior is evident in Figure 2.17 (a semilog y plot).

A detailed discussion of simple solar models has shown that, since the physics is understood, quite accurate models of stars can be made. They give confidence that the more accurate models than the ones shown here, give good representations of the structure of stars.

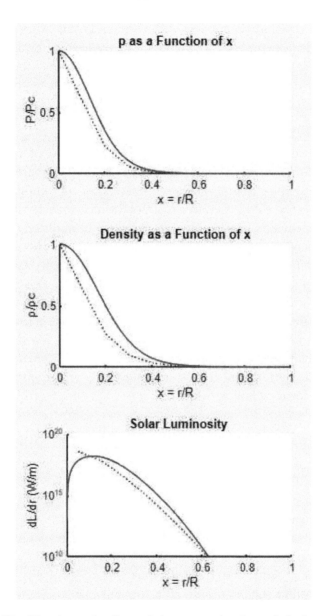

Figure 2.17:   The shapes for the scaled pressure density and the luminosity as functions of x.

Suffice it to say that the models of the Sun and other stars are secure in the hands of experts. The introductory treatments given here for simple ODE, polytropic, and BVP analyses should give the users an intuition as to the difficulties and advantages of each method.

2.10. Magnetic Bottle

A star can contain the fusion reactants by virtue of the stellar gravitational pressure. Fusion on Earth would be a great source of energy because the processes release MeV of binding energy compared to the eV of chemical reactions which reflects the relative scale of nuclear and atomic binding energies. In addition, there is a lot of hydrogen and deuterium fuel available on Earth. However, a hot plasma with temperatures of millions of degrees is needed comparable to or greater than the core temperatures of the Sun, so containment by any material vessel is out of the question. Containment of charged particles using magnetic fields in a magnetic "bottle" is possible and has been the subject of much research over the past few decades.

The fields outside the electric charges and currents obey Maxwell's equations in free space and only those are possible candidates for a confining field:

$$\vec{\nabla} \cdot \vec{B} = 0, \quad \vec{\nabla} x \vec{B} = 0 \tag{2.32}$$

A charged particle, charge q, with velocity v is affected by the Lorentz force. For many years, people have made sophisticated attempts to choose magnetic field configurations which contain the plasma of fusion reactants.

A simple set of magnetic fields is defined and particles are tracked in 3D position and velocity in the App "Tokamak_Bottle". The code uses the utility "ode45" to track the particle and the utility "plot3" is applied to follow the particle in all three dimensions. The user of the App chooses a set of three positions and three velocity components to set the initial conditions. Results for a specific choice are shown in Figure 2.18. In the specific example, over the time range plotted, the charged particle is reasonably well contained by the fields. The user can vary the inputs in order to see if the bottle "leaks" when the

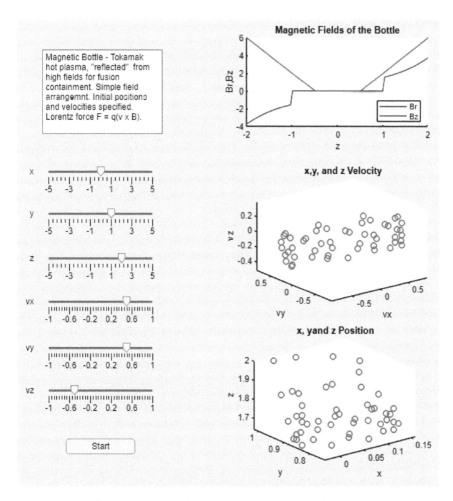

Figure 2.18: Output plots of the trajectory of a charged particle in a specific attempt at a "bottle" with the initial $x.y.z$ position and velocity vx, vy, vz chosen by Sliders. Over the plotted range, the positions of the particles are localized in space.

inputs are varied more widely. If a user is more ambitious, the actual magnetic fields can be changed by editing the instructions in the CodeView. Needless to say, much more clever field configurations have been tried, but fusion power is still in the future.

2.11. Fusion on Earth

There are several sources of energy that people have used historically. Any time a system is bound, energy can be released. For chemical reactions, the electrons are active, with energies released in the eV range. The waste products include many hazardous chemicals, such as CO_2. For bound nucleons, the energy scale is MeV, and the fission process was the earliest energy source. Fission is initiated by slow neutron capture in specific heavy nuclei such as U^{235}. These fuels are rare and expensive to produce. A typical process is $U^{235} + a \rightarrow Ba + Kr + 2n$ releasing about 200 MeV of kinetic energy per fission or 8×10^{13} J/kg of fuel which fissions. This energy release is a great improvement, but the reaction products are highly radioactive and long lived and they constitute a major issue for this energy source. The final state neutrons also need to be slowed down in order to make the precess self-sustaining because the neutron capture process requires a slow neutron to be efficient. For these reasons, fusion seems a likely better candidate with few waste products and abundant fuel.

Attempts to make a fusion reactor on Earth have been varied. A multibillion dollar effort to focus many lasers onto a small fuel pellet and implode it so as to raise the pellet to high temperatures has been one approach, called "inertial confinement". However, by far the most popular approach, at least so far, has been magnetically confined fusion. However, the ionized plasma is confined and the goal has been a "break-even" situation where more energy is released in fusion than must be supplied to operate the "reactor". In what follows, it will simply be assumed that a magnetically confined region exists at an elevated temperature and that region contains deuterium as a fuel. Indeed, deuterium occurs naturally and can be extracted from sea water (heavy water).

The probability density as a function of particle energy, ε, and temperature, T, is defined by the Boltzmann factor, suitably normalized. The dependence on energy and temperature has already been shown in Eq. (2.16). Because the densities are not enormous, Maxwell–Boltzmann statistics can be used. The units for $dn/d\varepsilon$ are inverse energy and inverse volume.

The number density, $dn/d\varepsilon$ is the probability of finding an energy ε between ε and $\varepsilon + d\varepsilon$ normalized to the total number density. The reaction rate density, Γ/V, depends on the deuterium velocity, v_D, the number density, n_D, and the D–D fusion cross section, σ_{DD}. The number density enters as the square because there are two deuterons needed per reaction so that the joint probability is required, i.e.,

$$dn/d\varepsilon = 2n\sqrt{\varepsilon}[e^{-\varepsilon/kT}]/[\sqrt{\pi}(kT)^{3/2}]$$

$$d(\Gamma/V)/d\varepsilon = n_D^2 v_D \sigma_{DD}[dn(D)/nd\varepsilon], v_D = \sqrt{2\varepsilon/mc^2} \qquad (2.33)$$

$$= [2c\sqrt{2/\pi}]n_D^2\sigma_{DD}[\varepsilon e^{-\varepsilon/kT}]/[(kT)^{3/2}\sqrt{m_D c^2}]$$

The D–D cross section itself is very temperature dependent because the Coulomb repulsion of the reactants must be overcome. Indeed, that is why the core of the Sun where fusion occurs exists near the center of the Sun where the temperatures are highest, as has been shown previously. The cross section rises exponentially with the square root of the energy. The common unit used in nuclear physics for the cross section is the barn, $1b = 10^{-28}m^2$. The commonly used energy unit is MeV $= 10^6 \times 1.6 \times 10^{-19}$J:

$$\sigma_{DD} = a(e^{-b/\sqrt{\varepsilon}})/\varepsilon \qquad (2.34)$$

The energetics of a fusion reactor is explored using the App, "DD_Fusion". The core of the Sun has a temperature of about 15 million degrees and a density about 150 times that of water. Approximately similar values would be needed in a functioning fusion "reactor", but those densities cannot be achieved using a charged plasma of nuclei because the dense aggregation of charged particles would repel one another. Reduced volumes and densities with respect to the Sun need to be compensated for with increased temperatures.

The figure created by the App is shown, in Figure 2.19. The Slider units for number density used here, $10^{22}/m^3$, are not comparable to Avagadros number of $N_A = 6.02 \times 10^{26}/(kg^*mol)$ or the core number density of the Sun, about $10^{32}/m^3$. One mol is N_A particles, where N_A is the number of atoms needed to make an atomic weight, A, in kg or $1/m_p$ times density units in kg/m^3, $\sim 6 \times 10^{26}$.

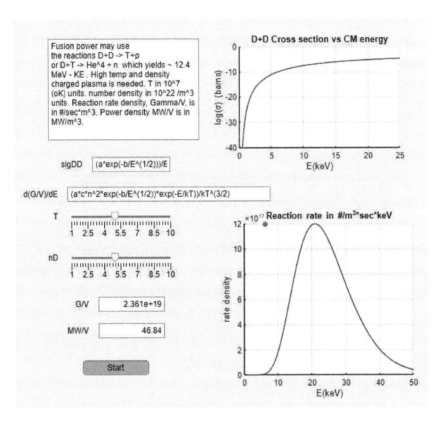

Figure 2.19: Figure created by the App "DD_Fusion". The cross section and reaction rate density as a function of energy are plotted.

The symbol for particle energy is E in the App and not ε because that symbol is not available to the App Text box. The red star in the reaction rate plot is the mean thermal energy, $\langle \varepsilon \rangle = 3kT/2$. The fact that the D–D reaction rate is very energy dependent skews the peak of the fusion power to higher energy values. The rate density, Γ/V, and power density, W/V, are in units of sec^{-1}m^{-3} and MW/m^3, respectively. The Matlab utility "quad" is used to integrate $d(\Gamma/V)/d\varepsilon$ over energy to find the rate density. The power density follows since 12.4 MeV of kinetic energy per fusion is released, on average. Because the cross section increases by 30 orders of magnitude in the figure, high temperature can overcome the advantages that the Sun has in number density and volume.

The user can vary the temperature of the plasma and the number density of the reactants in order to get an intuition on the dependence of the power density on these quantities. As expected, temperature is the crucial factor. As seen from the figure, a device with $20\,\text{m}^3$ of the specific reactant at a temperature of about four times that of the solar core would yield a fusion power of about 1,000 MW, neglecting the input power needed to operate the reactor. This is the allure of fusion research.

Indeed, a protype reactor designed to achieve break-even operation, when net power is produced, is now being constructed by a large international consortium. A schematic representation of the "ITER" or International Thermonuclear Experimental Reactor is shown in Figure 2.20. The size is immense as is the cost. There are many practical issues not mentioned in this simple fusion discussion. To choose one, the neutrons produced in the fusion reactions are not contained by the magnetic fields and their energy must be dissipated in the materials of the fusion reactor itself.

Figure 2.20: Schematic representation of the ITER reactor. The magnetic confinement coils are shown in orange. The confined plasma appears as a shining volume inside the coils.

As the cynics say, "fusion as an energy source is only 20 years away" and "the statement is always true".

Chapter 3

Stellar Evolution

"We are made out of stardust. The iron in the hemoglobin molecules in the blood in your right hand came from a star that blew up 8 billion years ago. The iron in your left hand came from another star."

— **Jill Tarter**

"The black holes of nature are the most perfect macroscopic objects there are in the universe: the only elements in their construction are our concepts of space and time."

— **Subrahmanyan Chandrasekhar**

"The ultimate extreme level of a moment in time is the singularity of a black hole."

— **Khalid Masood**

3.1. Stellar Pulsations

Previously, an estimate of the stable lifetime of a typical star like the Sun was found to be several billion years before the materials for fusion were exhausted. All good things must come to an end, and it is time to consider the endpoints of stellar evolution. The force of gravity is always attractive. The electrical forces can be repulsive, but macroscopic objects appear to be electrically neutral. Pressure stabilizes gravitational collapse, but the pressure is generated by hot fusion reactions, which must cease as fusion reaches a point where sufficient binding energy can no longer be released.

All stars are finally unstable. As the primary H fuel in the core depletes, a residual He core remains as the deeply bound "cinders". The inert helium core is surrounded by a hydrogen shell, where primary fusion continues. The core contraction releases gravitational energy and the envelope can expand and cool. An example star is Arcturus, a nearby "red giant" which has expanded and cooled shifting its spectrum into the red end.

Convection will dominate in the expanded interior. As helium is created and hydrogen is depleted, fusion switches to C, O reactions which are less efficient for energy production and hence the star cools and expands. Later, as these fusion rates slow, the outer layers may be ejected and a white dwarf remains in most cases. In this case, the white dwarf is an old red giant with a degenerate C–O core. A schematic diagram of a possible stellar evolution path is shown in Figure 3.1.

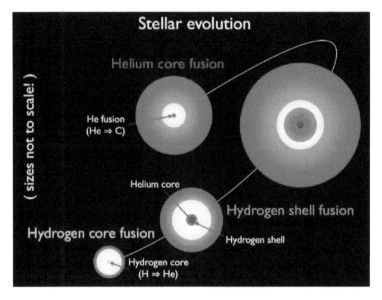

Figure 3.1: Schematic diagram of a branch stellar evolution. The star initially is largely composed of hydrogen which fuses in the core into helium. As the core hydrogen is used up, fusion continues in a hydrogen shell outside the helium core. At some point, the helium core begins to fuse to heavier elements such as carbon, nitrogen, and oxygen.

A red giant may have a mass of about 0.2–8 solar masses with a surface temperature about half that of the Sun. However, the radius is approximately 10–100 times that of the Sun. A typical red giant has a C, O core with helium and an outer shell where H is fusing followed by an outer hydrogen–helium envelope. As the star contracts, it will become either a white dwarf, stabilized by the Fermi pressure of electrons or neutrons, or it will, if massive enough, explode into a supernova. These are quite different end points, and they illustrate the competition between gravity and pressure.

Some stars are observed to oscillate. These classical Cepheids are variable stars somewhat more massive than the Sun and are more luminous. They are typically yellow or red giant stars. Their radii may change substantially during a pulsation cycle. Cepheids are used to determine distances to nearby galaxies because there is a very tight relationship between their luminosity and their oscillation period, which may vary from hours to years. Thus, they are "standard candles" for nearby galaxies. To be useful, the star must be individually visible which limits their use as distance markers to nearby, resolvable, stars. For larger distances, Type I supernovas are used as "standard candles" as will be mentioned later.

What determines the oscillation pattern of such stars? One way to think of it is that the tenuous outer matter is affected by sound waves from the interior. This behavior is in analogy to the previous treatment of the possible collapse of a gas cloud in forming a protostar. The sound waves propagating through the body of the star, of radius R and density ρ, assuming constant density, have a sound velocity of v_s. The pressure waves, which depend on density and pressure, travel a distance R in a time, τ. The pulsation period goes as the inverse square root of the density, which can be correlated with the luminosity. That correlation allows the Cepheids to be used as nearby standard distance candles.

Assuming adiabatic processes with $\gamma = 5/3$ as was done previously, the period for a constant density star, using $P(r) \sim G\rho^2(R^2 - r^2)$ is given as follows:

$$v_s^2 = (\gamma P)/\rho$$

$$\tau = 2 \int_0^R dr/v_s \sim \sqrt{3\pi/(2\gamma G\rho)}$$

(3.1)

In order to explore the pulsations in more detail, a simple model is made. An outer mass shell is not in equilibrium and can oscillate. The basic equations are nonlinear for this dynamic situation as opposed to a previous study of stable states. In equilibrium, Eq. (2.22), the acceleration is zero.

$$\rho(r)d^2r/dt^2 = -GM(r)\rho(r)/r^2 + dP(r)/dr \qquad (3.2)$$

Simplifying the model to a central mass, M_c, a shell with constant pressure P_s at radius r_s with a mass m_s, the equations become more tractable

$$d^2r/d^2t = -(GM_c)/r^2 + (4\pi P_s r_s^5)/(m_s r^3) \qquad (3.3)$$

There can be oscillations depending on the relative magnitude of the two terms on the right-hand side of Eq. (3.3), or the shell can escape. There are two competing effects leading to oscillations. The shell may expand and cool thus causing less pressure so that the shell contracts which causes increased heating.

The pulsations of such a simple model of the star are shown using the App script "Red_Giant". Values of M_c, and m_s are fixed. The user chooses the shell pressure by a Slider. In equilibrium, there is no acceleration, $GM_c\ r_{eq} = 4\pi\ P_s\ r_s^5/m_s$. The output of the App is shown in Figure 3.2. The user can choose a low enough shell pressure, using the Slider provided, that the shell oscillation becomes unstable and begins to diverge. The solution is obtained by using the Matlab utility "ode45". The equilibrium radius is calculated and displayed on a Linear Gauge.

The differential equation and the equilibrium radius are displayed symbolically using an EditField text with the "char" utility to display symbolic variables as text strings. Although the equations are nonlinear, the solutions are approximately those of simple harmonic motion (SHM) when the motion is stable. The scale of the period is a fraction of a year for the defined choices of central mass and shell mass.

Data from an oscillating star in the constellation Cygnus are shown in Figure 3.3. The period is about 1 year in this case. As seen in Eq. (3.1), the period is expected to scale as the inverse square root of the density. High-density stars will have shorter oscillation periods. This type of behavior was already mentioned in the context

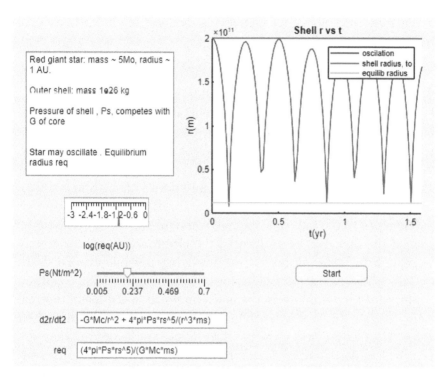

Figure 3.2: Output figure of the App "Red_Giant". The pulsations of the shell are plotted for the chosen shell pressure.

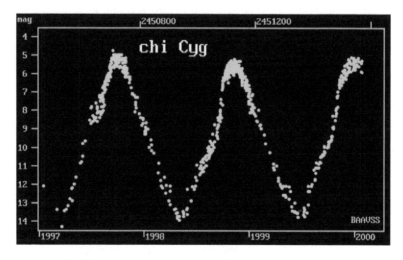

Figure 3.3: Oscillation data from a star in Cygnus. The period is about 1 year.

of the time for collapse of gas clouds, leading to the formation of protostars.

3.2. Pressure

As stars burn out and cool, they contract under gravity and are perhaps then stabilized by the Fermi pressure of the electrons. The hydrogen in the core remains ionized. For a 13.6 eV binding energy, the corresponding temperature is about $1.6 \times 10^5 \, °\mathrm{K}$, which is about 10 times less than the current core temperature of the Sun.

About 90% of all stars form "white dwarfs" with radii approximately equal to the radius of the Earth and masses comparable to the mass of the Sun. For white dwarfs, the electrons are approximately "degenerate". That means all the low-lying possible energy states are full, while all states of higher energy are empty. That condition is fulfilled if the Fermi energy is much larger than the mean thermal energy $3kT/2$. In that case, the electrons are mostly not in excited states. Cold and dense systems of fermions are favored if they are to provide a degenerate system. The number density for electrons is approximately $(Z/A)(\rho/m_p)$.

The Fermi energy is, in the non-relativistic (NR) case, the square of the Fermi momentum divided by twice the particle mass. The value of the Fermi energy, ε_F, is set by the Fermi Exclusion Principle that one spin 1/2 particle, like the electron or the neutron, cannot occupy the same quantum state as another and that quantum states require, in a phase space defined by the Heisenberg uncertainty principle. The Fermi momentum, p_F, scales as the third power of the number density, which reflects the counting of quantum states. The wave number associated with p_F is k_F. The fermions remain NR as long as the Fermi momentum is approximately less than mc, where m is the fermion mass.

$$\varepsilon_F = (\hbar^2 k_F^2)/2m, \quad k_F = (3\pi^2 n)^{1/3}, \quad \beta_F = \hbar k_F/mc \qquad (3.4)$$

The energy of gravity is $U_G \sim GM^2/R \sim GM^2 \, V^{-1/3}$. The pressure is $P_G \sim dU_G/dV \sim GM^2/R^4 \sim GM^2/V^{4/3}$. In comparison, the Fermi energy is $\varepsilon_F \sim n^{2/3} \sim (N/V)^{2/3}$ or $U_F \sim N\varepsilon_F \sim N^{5/3}/V^{2/3}$. The Fermi pressure is $P_F \sim dU_F/dV \sim (N/V)^{5/3}$. The number density scaling for the Fermi pressure is faster than that for gravity as long as the fermions remain NR. If they become UR, the scaling softens from a 5/3 power to a 4/3 power because $\varepsilon_F \sim cp_F$, and the balance becomes unstable. The pressure of gravity compared to the Fermi pressure more exactly, with the correct numerical constants shown for a constant-density star, is given as

$$P_G = (3/8\pi)(GM^2/R^4)$$
$$P_F = (3n/2\pi)^{5/3}(\hbar^2\pi^3)/(15m) \qquad (3.5)$$
$$\beta = \hbar(3\pi^2 n)^{1/3}/(mc)$$

For example, silicon at room temperature has a Fermi energy $\sim 3\,\mathrm{eV}$, and at room temperature, $kT \sim 0.025\,\mathrm{eV}$, most electrons are not thermally excited. For the Sun the Fermi energy is $\sim 4.5\,\mathrm{MeV}$ at the core, while kT_c is $\sim 1.25\,\mathrm{keV}$ and the electron pressure is only about a factor 0.027 of the core pressure. The electrons in the core are NR, with a value of $\beta \sim 0.042$. However, the C–O white dwarf Sirius B has a Fermi pressure of approximately 10^{22} Pa about a million times higher than the core pressure of the Sun. Note, that the fermions are assumed to be NR. The NR electron gas is a polytrope with index γ of 5/3 while the UR gas has an index of 4/3.

If the electron pressure stabilizes the star, with M less than about 1.5 solar masses, a white dwarf can form. Most stars in the galaxy will end up as white dwarfs. The equilibrium white dwarf radius, assuming NR electron balance, is $P_G \sim GM^2/R^4 = P_F \sim (N/R^3)^{5/3}/m$ or $R \sim N^{5/3}/(GM^2 m)$. In detail,

$$R = (9N/8\pi^2)^{5/3}(8\pi/3GM^2)(\hbar^2\pi^3/15m) \qquad (3.6)$$

If the electron pressure is insufficient, the star will continue to contract, the electrons will be forced into the protons and a neutron star may form. The process $p + e^- \rightarrow n + v_e$ yields an energy release of $0.78\,\mathrm{MeV}$. Using Eq. (3.4), an estimate of the densities when the electrons are relativistic, $n_e = n_p \sim (m_e c/h)^3 \sim 10^{37}\mathrm{m}^{-3}$ and $P_F \sim n^{5/3}h^2/m_e \sim 10^{24}$ Pa. However, when the neutrons become relativistic, the stellar radius is very close to the Schwarzschild radius, $R_s = 2GM/c^2$, and a black hole singularity may form depending on the exact details. In this text, the radius R_s can be thought of a distance where gravity becomes a strong force.

The stabilizing effect of Fermi pressure is illustrated using the App "Fermi_Press_e_n". The plot shown in Figure 3.4 illustrates the radius and pressure for a star of mass chosen by the Slider. A uniform density is assumed. The black point is the Schwarzschild radius. The triangle shows the, location of the Sun on the plot. The electron and neutron pressures are only plotted when p_F/mc is less than 1. The approximation used here is that for smaller Fermi momentum, an NR dependence is used, while for higher momenta, the Fermi pressure is wholly ineffective. The point where this occurs is displayed numerically as R_e and R_n in m. The ratio is approximately the electron to neutron mass ratio. The Schwarzschild radius is shown as R_s in km.

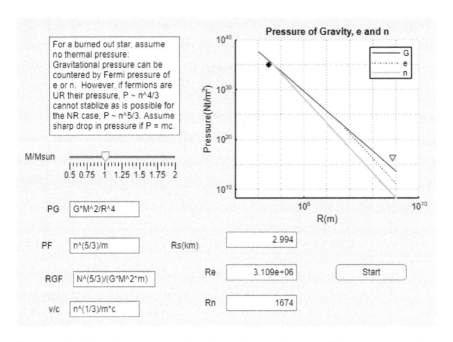

Figure 3.4:   Plot of pressure as a function of radius for a specifically chosen stellar mass value. The radii where the electrons and neutrons become UR are displayed numerically, and their points are suppressed at higher pressures in the plot.

Symbolic dependencies for the gravitational and Fermi pressure and the equilibrium radius are shown using the "char' utility for text strings, as are the numerical values. Stars with masses approximately that of the Sun may be unstable even with neutron pressure as can be seen by varying the mass Slider. In the plot, as R decreases, the electron and neutron pressures rise, while they are NR, faster than the gravitational pressure. Where the lines intersect, if the point is before UR behavior, the star is stabilized.

3.3. White Dwarf

It appears that there are three possible end points for a star. It can be electron stabilized and form a white dwarf. If it is too massive for that, it can collapse to a smaller and denser neutron star. If that also fails, the formation of a black hole is the endpoint, a situation where nothing can stabilize the object from being gravitationally compressed into a "point". As seen in Figure 3.4, for a star of one solar mass, the electron-stabilized white dwarf that forms would have a radius of order that of the Earth.

Most of the stars on the main sequence end up as "white dwarfs" stabilized by electron Fermi pressure. Heavier stars may become supernovae or neutron stars. A white dwarf has a central pressure given approximately as $P = G\rho^2 R^2$ or 6×10^{22} Pa, which is about a million times greater than the core pressure of the Sun.

The parameter x in Eq. (3.7) defines the NR behavior of the electrons. The number density n_o is reached when the electron momentum is $m_e c$ and follows from Eq. (3.5). The mass density is defined by the nucleons in the gas and is assumed to be largely iron (binding energy curve peak) with $Ye = 0.46$, which reflects the approximately equal number of protons and neutrons for iron. The electron number density increases with x, as the electrons occupy higher momentum quantum states:

$$n_o = (m_e c/\hbar)^3/3\pi^2$$
$$x = p_F/(m_e c) = (n/n_0)^{1/3}$$
$$\gamma_F \sim x^2/\sqrt{1 + x^2}$$
$$\rho = (m_p n_o)/Y_e$$

(3.7)

Previously, the NR to UR transition was assumed to be abrupt. A smooth behavior is now used to scale the behavior as the UR limit is reached by introducing the function γ_F which gives a goes as $\sim x^2$ at small x and $\sim x$ at large x so that the energy has proper NR and UR behavior, with energy proportional to momentum squared for NR but proportional to momentum for UR. In this way the binary behavior, either NR or UR and thus offering no Fermi pressure in the latter case, is smoothed out.

Since the gravitational pressure is given as GM^2/R^4, while the Fermi pressure scales as $(M/R^3)^{5/3}$, the scaling equilibrium is approximately that MR^3 is a constant. More massive white dwarfs are, perhaps surprisingly, smaller. From the prior discussion, Eq. (3.6), an equilibrium variable, displayed in Eq. (3.6), is defined to be R_o, Eq. (3.8). It is used as a scaled variable, with an assumed constant density. This NR treatment is now made slightly more accurate as the UR regime is approached:

$$R_o^2 \sim (n_o^{5/3}\hbar^2)/(m_e G \rho_o^2)$$
$$M_o = (4\pi/3)R_o^3 \tag{3.8}$$

The code "White_Dwarf_App" is employed to look at the situation specifically when the electrons are becoming relativistic. The Figure generated by the App is shown in Figure 3.5. The electron density when the behavior is UR, n_o, is shown numerically as is the corresponding nucleon mass density. The equations which are solved for numerically are as follows:

$$dM(r)/dr \sim r^2\rho(r)$$
$$d\rho(r)/dr \sim -[M(r)\rho(r)]/(\gamma_F r^2)] \tag{3.9}$$

The equation for dM/dr follows from Eq. (2.22), while $d\rho/dr$ follows from dP/dr, modified by γ_F to account for the relativistic transition. The scaled variables R_o and M_o are also calculated and displayed. Solutions are found when the radius is about 1% of the radius of the Sun while the mass is slightly larger than that of the Sun. Electron pressure stabilizes the radius until the radius begins

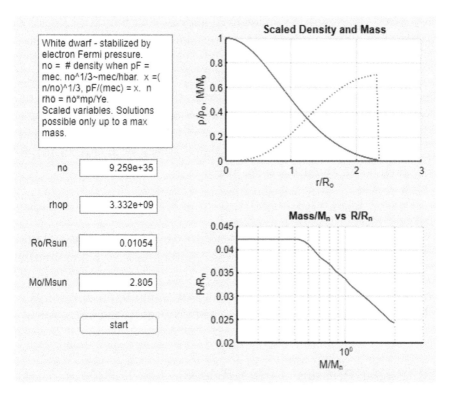

Figure 3.5:   Plot of the behavior of the scaled density and mass as a function of the scaled radius. The second plot is the implicit behavior of the radius as a function of the mass of the white dwarf.

to fall with mass as the electrons become UR. The plots terminate when a stable solution cannot be found. The density falls with r and the mass increases. As the mass increases, the radius falls as expected from qualitative arguments made previously. The Matlab utility "ode45" is used to solve the approximate differential equations for density and mass, where the mass is due to the nucleons, but the pressure is due to the electrons.

As can be seen, there is a falloff of the density with radius. However, the scaled value of M_o is not reached because the solutions become impossible. This treatment is very approximate even though an attempt is made to follow the electrons as they become UR and the density is no longer uniform. Many detailed calculations have been made. The resulting famous "Chandrasekhar limit" is 1.44 solar masses, in somewhat fortuitous agreement with the present crude estimation.

Any star with a mass above that limit must continue to collapse. The maximum mass of a white dwarf that can be stabilized by electrion Fermi pressure can be estimated another way. Equating the core pressure for a constant-density star to the Fermi pressure and assuming constant density, the maximum stellar mass, Eq. (3.10), is estimated to be about 0.4 solar masses:

$$M_{\max} = (3\sqrt{2\pi}/8)(\hbar c/G)^{3/2}[(Z/A)(1/m_p)]^2 \qquad (3.10)$$

More accurate calculations lead to a maximum mass about 1.5 solar masses depending on the approximations made in the calculation. It is of note that no white dwarf having a higher mass has been found. A solar mass neutron star would have only about a 3-km radius because the neutrons remain NR up to a higher pressure and such a star would contain about 10^{57} neutrons.

3.4. Stellar Pressure — Relativistic

Before leaving the topic of white dwarf stars, there is another factor that has been ignored because it is small for the typical star but which could be dominant for a neutron star. For example, the Sun has a core pressure of only about 1.9×10^{-6} of the core density time c squared. However, in GR, all energy has gravitational consequences so that pressure contributes to the energy density. For a uniform-density star, a GR interior closed-form solution exists and the pressure contribution can dominate.

The Schwarzschild radius, R_s, as will be shown later, is a crucial parameter and may be considered the radius of a black hole. For now it can be thought of as a distance scale when the gravitational energy, GM^2/R_s, is comparable to the rest energy Mc^2. The gravitational potential $V = M\Phi$ has the dimensions of velocity squared, and with $R_s = 2GM/c^2$, $\Phi/c^2 = R_s/2r$ is dimensionless. Therefore, for $r \gg R_s$, the potential energy is much less than the rest mass, and the system can be considered to be well described classically.

The parameter $x = P/(\rho c^2)$ indicates the importance of the effect of pressure. Classically, the interior of a uniform-density, or incompressible, star of radius R has a derivative dx/dr which is negative and increases with r. The solution for $x(r)$ is shown in Eq. (3.11) for the NR or Newtonian case. It is notable, perhaps, that the results can be expressed as dimensionless ratios of powers of r and R_s. In fact, the energy–momentum tensor for a fluid at rest is diagonal with temporal term $= \rho$ and spatial term $= P/c^2$. The pressure is a maximum at the center and vanishes at the surface, $x(0) = R_s/4R$, $x(R) = 0$, i.e.,

$$x = P/(\rho c^2), \quad R_s = 2GM/c^2$$
$$(dx/dr)_N = -(R_s r)/2R^3, \quad x = [R_s/R - (R_s r^2)/R^3]/4 \tag{3.11}$$

In the GR case, the source of the metric tensor is the stress energy tensor. It has contributions from both mass and pressure as might be expected from examination of Eq. (3.11). The behavior of the GR interior solution is more complex, but approaches the Newtonian case for a constant-density stellar interior when r is much greater than R_s. There is a closed-form solution in this simple case, which appears in Eq. (3.12). The interior metric is continuous with the exterior Schwarzschild metric at $r = R$ and $P(R) = 0$. The central pressure is $P(0)/\rho c^2 \sim R_s/4R$, at small R/R_s, which indicates that pressure is important only for radii near to R_s. For more complicated densities, an equation of state, such as $p \sim \rho^\gamma$, of the polytropic form, can be invoked to solve for the density numerically as is done classically:

$$ds_{\text{int}}^2 = [(3/2)\sqrt{1 - R_s/R} - (1/2)\sqrt{1 - (R_s r^2/R^3)}]^2 (cdt)^2$$
$$-dr^2/(1 - R_s r^2/R^3) + r^2 d\Omega^2$$
$$(dx/dr)_{GR} = -((R_s/2r^2)(m(r)/M)(1 + x)[1 + 3x])$$
$$/[1 - (R_s/r)(m(r)/M)]$$
$$P_{GR}/\rho_o c^2 = -[\sqrt{1 - (R_s r^2)/R^3} - \sqrt{1 - (R_s/R)}]$$
$$/[3\sqrt{1 - (R_s/R)} - \sqrt{1 - (R_s r^2)/R^3}] \tag{3.12}$$

The mass $M(r)$ is the interior mass at r. The mass $M(R)$ is the exterior mass, M, by continuity. Classically, when matter is packed under its self-gravity, there is a binding energy loss of $(\Delta M)/M =$

$(3/10)(R_s/R)$. In GR, M refers to all mass energy in the source, even the negative binding energy. When a proper volume element is considered in GR, the proper volume times ρ is not M but has a mass defect equal to the binding energy$/c^2$.

The classical limit is $dP/dr = -4\pi G\rho^2 r/3$, which agrees with Eq. (2.14). The interior and exterior solutions for the Newtonian potential are given for reference. There is a smooth matching of the interior and exterior potentials:

$$\Phi_{\text{int}}/c^2 = (R_s/4R^3)(r^2 - 3R^2)$$
$$\Phi_{\text{ext}}/c^2 = -R_s/(2r)$$

$$(3.13)$$

The Newtonian potential and the Newtonian and relativistic values for the interior pressure of a uniform star are displayed using

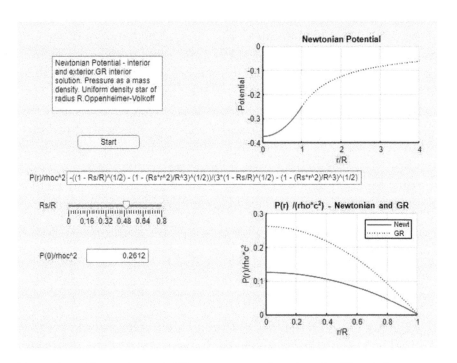

Figure 3.6: Newtonian and GR pressure as a function of $r/R < 1$ for a user-chosen choice of R_s/R. In this particular case, the effect of pressure and density are about equal.

the App "Star_Interior_Press_GR" and are shown in Figure 3.6. There
is no GR "potential" because that is an NR concept which assumes
action at a distance, while in GR, actions move at the speed of light,
at most. In a Newtonian limit however, the potential can be identified
with the diagonal time component of the energy–momentum tensor.
The user chooses a value of R_s/R using a Slider. The symbolic
expression for $P(r)$ is displayed for reference.

For the Sun, the value of R_s/R is about 4×10^{-6}, so a Newtonian
treatment is valid. Using the Slider it is easy to see that for small
values of R_s/R the pressure term is small while for large values the
ratio $P/(\rho_0 c^2)$ is greater than 1, indicating strong relativistic effects.
Pressure indeed aids the collapse. For white dwarfs and neutron stars,
these effects cannot be ignored. Realistic models for endpoint stars
may well need to take relativistic effects into account since the radii
in question may not be small with respect to the Schwarzschild radius
and gravity has become a strong force.

3.5. White Dwarf — Gravitational Radiation

A white dwarf is an object of extreme density, which then may be
a source of GR radiation. This radiation is quadrupole as opposed
to the EM dipole, because there is no negative mass and therefore
no mass dipole. The radiated power therefore goes as the frequency
as ω^6 rather than the fourth power familiar from the case of the
EM dipole. Any accelerated mass quadrupole radiates gravitational
waves which propagate at the speed of light. However, in order to be
measurable, the radiating system needs to be extreme, with distance
scales of order R_s.

There is a maximum possible power radiated by gravitational
systems in extreme circumstances, $W_{\max} = (2c^5/5G)$, which is 1.33×10^{52} W. This is approximately 10^{26} times the solar luminosity which
indicates that degenerate stars are fundamentally different from main
sequence stars. These types of stars appear to be responsible for some
of the most dramatic events observed by astronomers.

In general, a binary system rotates about its center of momentum
(CM), which is fixed here at the origin. Masses are M_1 and M_2 and

their distances from the fixed CM at (0,0) are r_1 and r_2 oppositely directed. Their velocities about the CM are $v_1 = \omega r_1$ and $v_2 = \omega r_2$. Consider the gravitational radiation emitted by a white dwarf binary system as it spirals to collision. Take for a limit in Eq. (3.14) the simplest numerical case where the masses are equal to M and their separation is $2R$ as they orbit about their common CM. The frequency of the orbit is then $\omega^2 = GM/(4R^3) = (2\pi/\tau)^2$. The GR radiation has a frequency 2ω because the quadrupole is the same after a half period.

The radiated power in gravitational radiation reaches the maximum when $2R$ reaches the Schwarzschild radius, $R_s = (2GM)/c^2$. Therefore, a binary system of white dwarfs can, in principle, radiate gravity waves with a power equal to that of the photons emitted by 10^{26} suns if they collapse into a black hole. This is unlikely because $R \gg R_s$ for white dwarfs, so that they will collide well before reaching separations $\sim R_s$:

$$\omega^2 = G(M_1 + M_2)/(r_1 + r_2)^3 \rightarrow GM/(4R^3)$$

$$dU_G/dt = (32/5)(G^4/c^5)M_1^2 M_2^2(M_1 + M_2)/(r_1 + r_2)^5$$

$$\rightarrow (2c^5/5G)(R_s/2R)^5$$

$$\sim (R_s/R)^4(Mc^2)(c/R) \tag{3.14}$$

When the binary looses energy by radiation, the period, $\tau = 2\pi/\omega$, shortens. The change in the period, τ, with time is shown in general and for the special case of an equal-mass binary. The change in the period with time is simply, related to the ratio of the characteristic orbital distance $c\tau$ to R_s to a power 5/3 ignoring constants of order 1 or the ratio of R_s to R to a power 5/2. As the binary system approaches R_s, the period shrinks dramatically:

$$d\tau/dt = -(96/5)(4\pi^2)^{4/3}(G^{5/3}/c^5)(M_1 M_2)/[(M_1 + M_2)^{1/3}\tau^{5/3}]$$

$$d\tau/dt \sim (R_s/c\tau)^{5/3} \sim (R_s/R)^{5/2} \tag{3.15}$$

For an observer at a distance r away from an equal-mass binary system, a dimensionless distortion of the metric which defines the curvature of local space-time occurs with a dimensionless magnitude

h and a time-dependent phase $2\omega t$:

$$h = (8MGR^2\omega^2)/(c^4 r) = R_s^2/(2rR) \qquad (3.16)$$

The metric defines measurements of intervals of space and time, so that changes in the metric are observable using test masses and watching their relative motion. This behavior is in analogy to the motion of charges in a radio antenna.

These predicted values of radiated power as a function of the white dwarf mass and radius are examined in the App "White_Dwarf_GR_Rad" for typical values of the white dwarfs and are displayed in Figure 3.7. Because the typical radii are about 1% of the solar radius and the typical mass is of order the solar mass, the Schwarzschild radii are of order km while the stellar radii are of order 10,000 km. Thus, the GR radiation does not cause a rapid decay of the binary orbits, and that effect can be largely ignored.

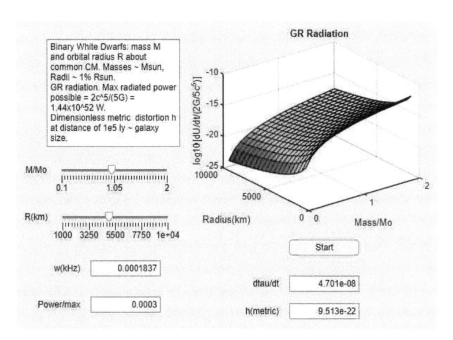

Figure 3.7: The resulting App figure for gravitational radiation from a white dwarf binary system.

The user can choose the white dwarf mass and radius using the Sliders provided. The frequency of the radiation is of order 0.1 Hz. The power radiated is of order a factor at least 10^{-4} less than the maximum possible power. The metrical distortion parameter, h, is of order 10^{-21} and the fractional change of the period per period is of order 10^{-8} therefore justifying ignoring the change in period here. More extreme stellar objects are expected to show the signs of the emission of GR radiation more clearly. The neutron star, stabilized by neutron Fermi pressure, is a more compact object by a factor of approximately 1,000.

3.6. Neutron Star, Pulsar

Fusion burns to iron nuclei as an end point since they have the maximum binding energy per nucleon. For a very massive star, the neutron Fermi pressure cannot stabilize the star, as was seen in Figure 3.4, and a supernova explodes. Successive shells where fusion has taken place end at an inner iron core followed by silicon, oxygen, carbon, helium, and an outer hydrogen envelope. The fusion rate slows and the core then collapses under gravitational pressure. A supernova results which may release a peak optical luminosity a few billion times that of the Sun.

Such bright supernovae can be observed at great distances and are used as "standard candles" by observing their characteristic light curves. The light curves show the spectral signatures of elements near iron on the periodic table, such as cobalt and nickel, as expected. The Cepheid variables are not bright enough to probe distances beyond our local galaxies. Using modern star surveys, many such supernovae have been observed. In fact, using them the existence of "dark energy" in cosmology was inferred leading to a Nobel Prize in 2011.

In 2017, a supernova in Cassiopeia A was observed and the spectra contained in the light curve showed the existence of heavy elements near iron. In fact, much of the periodic table has been populated by supernovae since with normal stellar evolution, the heavy elements cannot be formed. Indeed, the Sun is at least a second-generation star and the Earth has many heavy elements. Truly, we are stardust. A plot of the periodic table showing which elements come from primordial nucleosynthesis, which from white dwarf collapse, and which from supernovae are shown in Figure 3.8.

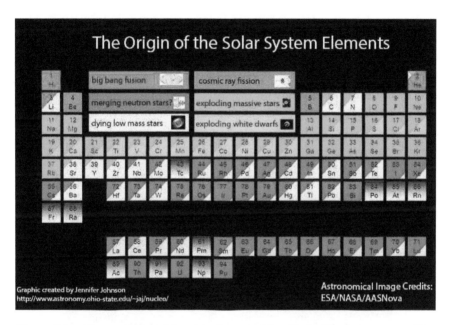

Figure 3.8: The periodic table with elements identified with their sources. The primordial elements are largely hydrogen and helium as discussed previously. The explosions of stars populate the elements near iron in the periodic table. The heaviest elements are thought to be formed by merging neutron stars.

The primordial elements are largely hydrogen and helium. Massive stars as they evolve burn heavier fuels to produce C, N, O, and other elements nearby in the periodic table. The majority of the elements above iron are ascribed to merging neutron stars or supernova explosions. These observations give strong support that there is a good current understanding of the end points of stellar evolution.

Turning now to the most extreme but still stable stellar object, the neutron stars are described. Under gravitational pressure, the protons and electrons are crushed into neutrons, releasing a lot of energy in the form of neutrinos. The reaction is $p + e^- \to n + v_e$ releasing 0.78 MeV of energy per neutron formed. If all 10^{57} nucleons in a typical neutron star were converted to neutrons, about 10^{44} J of neutrino energy would be released.

In 1987, neutrinos from supernovae were first observed on Earth in deep underground experiments from a source called SN1987. The optical detection of the event confirmed the supernova source of the underground neutrinos, which were detected in time coincidence. As the inner core collapses, the neutron Fermi energy stabilizes the core which rebounds and creates a pressure wave. This shock wave blows off the in-falling outer layers. Depending on specific details, the core could stabilize as a neutron star or form a "black hole". A 2017 Hubble image of SN1987 shows in Figure 3.9 the expanding ring of ejecta 30 years after the supernova explosion.

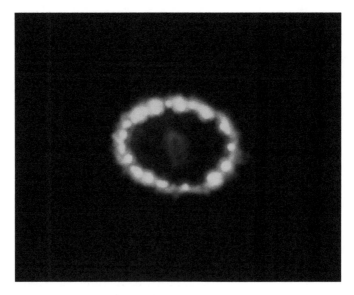

Figure 3.9: Hubble image from 2017 of the ejecta from the 1987 supernova explosion. The material has moved significantly away from the explosion.

For a neutron star remnant of mass M, the radius can be estimated by scaling the white dwarf radii. The estimated ratio of the white dwarf to the neutron star radius is simply the mass ratio of the neutron to the electron ~1880. The density for a 10-km neutron star could be roughly 10^{18}kg/m^3, which is a nuclear density. Another estimate would be to have a solar mass compressed to 5×10^{-6} of a solar radius or 3.5 km resulting in a density of 10^{19}kg/m^3. Note that these radii are near the Schwarzschild radius, 3 km for a solar mass, so that the neutrons barely stabilize the star. How might one observe a neutron star? It is small and no longer emitting visible radiation from fusion processes, so another method must be found.

One possibility is to use the X-ray portion of the electromagnetic spectrum. A neutron star binary or white dwarf binary are possible sources of X-rays. Recent data from X-ray observatories, the aptly named "Chandra" observatory for example, have opened up new tools to explore neutron stars.

A second possibility for observation leads to the idea of a "pulsar", which can be observable in the visible part of the spectrum. Neutron stars formed from spinning progenitor stars with magnetic fields will spin up to a faster rotational velocity since the moments of inertia scale as R^2. There will be stronger magnetic fields since magnetic flux is conserved and the surface area has shrunk drastically. A star with a typical field of order 10^{-5} T, seen in Zeeman splitting of solar spectral lines, would have a neutron star field of order 10^9 T. These fields are far beyond those that can be created on Earth. Electrons spiraling in these strong fields emit synchrotron radiation, which is detected on Earth as a pulse of radiation as the star rotates across the line of sight to the Earth. Thus, the neutron star is observable and, in fact, is seen to have a stable period, the rotation period. What is seen is called a "pulsar". The discovery of pulsars resulted in the 1974 award of the Nobel Prize. A schematic diagram of a pulsar and the magnetic fields is displayed in Figure 3.10.

Rotation also helps stabilize the neutron star because of the centrifugal repulsion which favors rapidly rotating neutron stars. The maximum angular velocity is set because gravity must hold the star together

$$\omega_{\text{max}}^2 R \sim GM/R^2 \qquad (3.17)$$

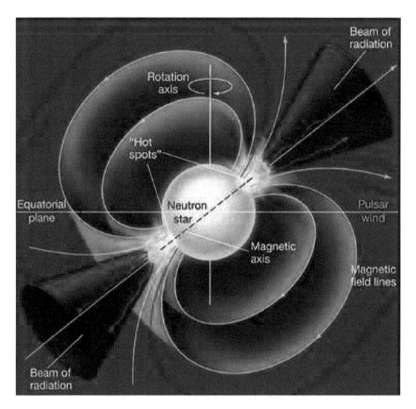

Figure 3.10: Schematic diagram of a pulsar. The neutron star rotates around its axis but the magnetic axis is not collinear. The magnetic field causes electron synchrotron radiation, which then appears to an observer as a pulsed beam of radiation with a stable period.

For example, a neutron star with a solar mass and a 10-km radius could sustain frequencies of up to about 10 kHz. The white dwarf frequency scale was seen to be Hz, Figure. 3.7. For neutron stars, the GR effects are much larger because the distance scales are about one thousand times smaller and can become comparable to the Schwarzschild radius.

A pulsar exists within the Crab nebula as a remnant of a past supernova explosion. The observation that a specific well-measured pulsar was slowing down at exactly the rate expected by general relativity due the radiated gravitational waves resulted in the award

of the Nobel Prize in 1993. The double-neutron star binary shows
the expected decrease in period.

The data are shown in Figure 3.11. The observed shift per orbit
is about 30 sec in 25 years or $d\tau/dt \sim 3.8 \times 10^{-8}$. The pulsar is indeed
an incredibly accurate clock but one which inevitably runs down as
it emits gravitational radiation. A comparable value of $d\tau/dt$ was
found for a white dwarf gravitational radiation in Figure 3.7.

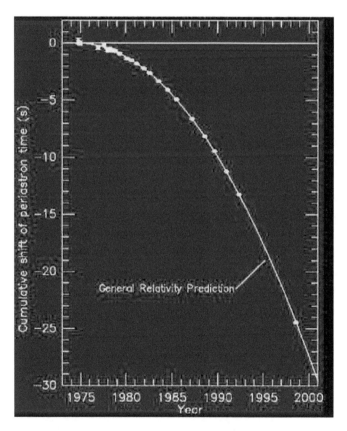

Figure 3.11: Data from 25 years of observing the period of a specific pulsar.
The curve is the decrease in period that is expected from GR. The agreement is
spectacular.

Magnetic dipole radiation from a compact neutron star is in the
X-ray region of the spectrum. The rate of slowing of the period allows

for an estimate of the magnetic field. Typically, pulsar fields of 10^8 T are inferred. The field at the pole is B and the angle is that between the field axis and the axis of rotation:

$$dU_{MD}/dt \sim (B^2 R^6 \omega^4 \sin^2 \theta)/(\mu_o c^3 \tau^4) \qquad (3.18)$$

For example, the Crab Nebula contains a pulsar that has a period of 33 ms and a slowing down of $d\tau/dt$ of 4.21×10^{-13} from which a field of 8×10^8 T is inferred.

3.7. The "Chirp"

For radii near R_s, the period cannot be taken to be constant or very slowly changing. In this situation, the emission of GR radiation causes the radius to shrink and the frequency to increase by noticeable amounts. The "chirp" is the characteristic increase in pitch as the binaries spiral into contact in the extreme case of collision, at least classically. A correct treatment requires full GR machinery which is not attempted here. However, some aspects are clear even in a Newtonian approximation.

At any radius the frequency $w(R)$ is shown in Eq. (3.19), $\omega^2(R) = (c^2(R_s))/(8R^3)$. The classical collision time, t_c, arises from integrating the decrease in period, Eq. (3.14), until the orbital radius vanishes for point-like masses. The fourth power of the radius is proportional to the difference between the time and the classical collision time, such that the binary radius is zero at t_c:

$$w(R) = (c/R)\sqrt{R_s/8R}$$
$$t_c = (5/4)[(R(0)/R_s)^3(R(0)/c)] \qquad (3.19)$$
$$R^4(t) = [(4R_s^3 c/5)(t_c - t)]$$

The results for a binary system are shown in Figure 3.12 using the App, "GR_Chirp_App". In this case, the integration is cut off at R_s since at that point classical physics is nonsensical. The initial values of neutron star mass and radius of the binary are set by user-chosen Sliders. A "movie" of the orbits of the two stars as a function of time is shown first. However, it is cut off when the radius is the Schwarzschild radius. The time to reach R_s is displayed and the

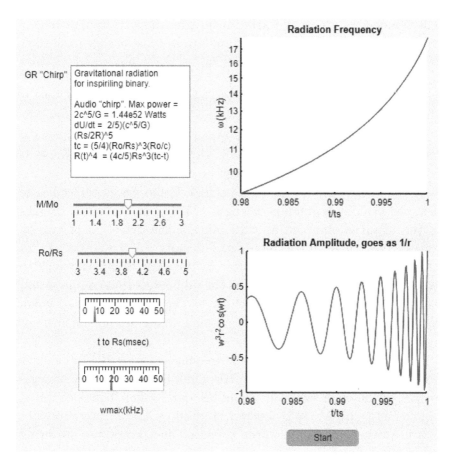

Figure 3.12: Results from the App "GR_Chirp_App" for a particular choice of initial neutron star mass and radius.

maximum frequency, which occurs when $R = R_s$, is also shown using LinearGauges for variety instead of the usual numeric EditFields.

The magnitude of the amplitude is displayed in a second figure. The characteristic increase in frequency that gives the name of "chirp" to the waveform is evident. A Matlab utility called "sound" is used to convert this waveform into an audible signal which drives the speakers of the user. The user can change the qualities of the audio and plotted chirp using the Sliders provided. The frequency as a function of t is displayed for times near the end of the "chirp".

The frequencies are in the kHz range. This analysis applies to any compact equal-mass binaries that radiate and lose energy until their Schwarzschild radius is reached, such as black holes and neutron stars.

There are many binary stars. Estimates are that perhaps 10% of all stars are binary. Many main sequence stars become white dwarfs. Since binary white dwarfs must gravitationally radiate, they will ultimately collapse. Therefore, it is expected that the Universe is full of gravitational waves that may be observable.

How would such waves be detected? Radio waves typically are detected in a dipole antenna of length $\lambda/4$. The motion of the charges in the metal is observed as a signal. For the detection of gravity waves a mass quadruple is required. The motion of the masses occurs because space–time itself is distorted by the wave. What is intrinsic to gravity and gravitational waves are the tidal forces that compress and elongate. Any uniform force, such as "g", can be cancelled by going to a free-fall reference frame. Hence, such forces are not intrinsic to gravity.

For binaries near their R_s, the frequencies are $\sim$10 kHz or have quarter wavelengths $\lambda/4$–50 km. For white dwarfs, with frequencies $\sim$1 Hz, the antenna size scale would be $\sim 5 \times 10^5$ km. That size would argue for a space-based detector. Indeed, a Laser Interferometer Space Antenna (LISA) has been proposed and prototypes are being deployed and tested.

3.8. Gravitational Antenna

The extremely small values for the metric distortion, (Figure 3.7 for example) have made direct observation of gravitational waves unobservable until technological advances have very recently enabled their detection. A simplified view of gravitational tidal forces, Figure 3.13, and the associated metrical distortions reflected in the motion of the "antenna" are given in the script "Grav_Rad_Tidal_3.m". It is the tidal forces that are intrinsic to gravity as mentioned previously. The tidal forces for a point source can be derived from a tidal potential. The x and z tidal potential and forces are, taking x to be transverse

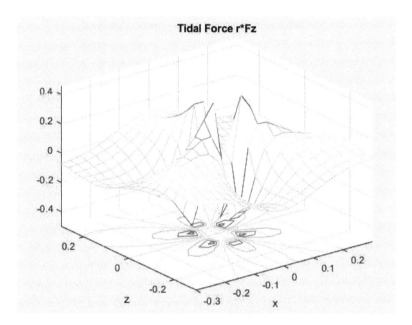

Figure 3.13: The tidal force in the z or axial direction. There is no force at the (x, z) origin since tidal forces only stretch and compress.

to the Newtonian force vector, given as follows:

$$\Phi_{\text{tide}} = -(R_s c^2/2)(z^2 - x^2/2)/(x^2 + z^2)^{3/2}$$
$$\vec{F}/m = -\vec{\nabla}\, \Phi_{\text{tide}}, \quad F_x/m = -(xR_s c^2)/r^3, \tag{3.20}$$
$$F_z/m = (zR_s c^2)/r^3$$

The force F_z times r is plotted in Figure 3.13. The scaling with r is simply to reduce the singular behavior that the force has at small r. The transverse force is compressive, while the longitudinal force acts to elongate a body of extended size.

There is a "movie" showing how an incident gravitational wave affects a test antenna, one that responds to the quadrupole in analogy to a dipole antenna for electromagnetic radiation. The last frame is shown in Figure 3.14. As the wave passes, the two transverse mass pairs are alternatively compressed and stretched. This behavior is a crude schematic of a real interferometer such as the ones used

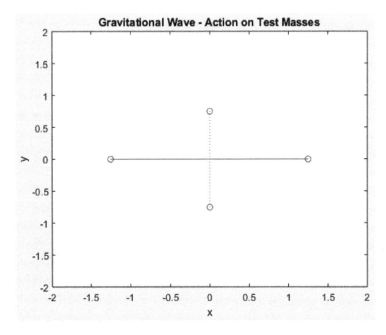

Figure 3.14: Action of a passing gravitational wave on a quadrupole mass antenna showing the stretching and compression of the two orthogonal transverse dimensions.

for gravitational wave detection but still displays the basics of the deformations.

The true apparatus is shown in Figure 3.15. It is essentially a giant interferometer, a scaled version of that used by Michelson and Morley but incomparably more sensitive. The two arms are in vacuum and the mirrors are carefully isolated from transient shocks.

Data taken in 2015 are shown in Figure 3.16 where a convincing simultaneous observation of a "chirp" at two independent observatories is shown. The strain on the interferometer, $\sim h$, is about 10^{-21} at maximum (see Figure 3.7). The period is shrinking, but is about $0.01\,$s or about $100\,$Hz. The interpretation could be that the signal is likely due to the merger of two black holes, with masses of about 30 solar masses. The detailed shape of the chirp yields information on the identities and properties of the two bodies forming the binary system. The treatment given here is much too simplistic to conclude

Figure 3.15: Picture of the Laser Interferometer Gravity-wave Observatory (LIGO) showing the building where the light is split and recombined after traversing the two orthogonal vacuum tubes.

anything about the binary system. The observation of gravitational waves for the first time was awarded the Nobel Prize in physics for 2017. The LIGO will continue to operate and upgrade. However, it has a low-frequency cutoff due to geologic noise and tremors. To overcome this problems an observatory in space, LISA, is proposed. While LIGO is sensitive to frequencies from 10 Hz to 1 kHz, LISA would look below 1 Hz. It would have a longer lever arm and not be sensitive to Earth tremors which should result in more sensitivity at lower frequencies.

3.9. Radial and Circular Geodesics

Although it is a bit premature, GR has been mentioned previously in discussion of stellar evolution end points so that thinking about orbits around black holes is now an appropriate topic for discussion. The expressions in this text, where relevant, are given in terms of the characteristic length of GR, the Schwarzschild radius. The weakness of gravity for most applications occurs because the radius, where gravity becomes strong, is so small.

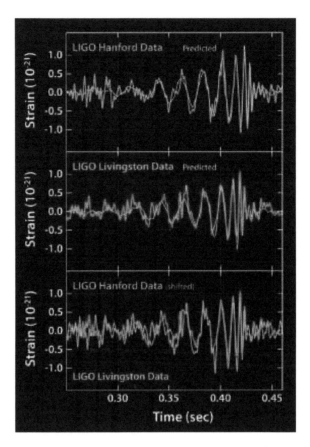

Figure 3.16: Data from two independent observatories of the characteristic "chirp" signal with increasing amplitude and decreasing period. These data show the discovery of gravity waves, opening a new way to look at the Universe beyond the electromagnetic spectrum.

Navigation near a black hole is a bit more complex than that near a normal star, which is what we will assume in Chapters 4 and 5. Even if we do not get to explore black holes directly, we can observe their effects on orbiting "test bodies" such as suns orbiting a black hole or remote probe missions and from them infer the properties, mass, and rotation, of the black holes that are being studied.

The basics of SR are useful to begin the discussion. The fundamental assertion of SR is that light travels at speed c in all reference frames moving in uniform relative motion, or "inertial frames".

In geometry, there is an invariant distance. The expanded version of distance is the invariant interval between two events in space–time. The interval, ds, has both temporal and spatial contributions:

$$ds^2 = (cdt)^2 - dr^2 - r^2(d\Omega)^2$$
$$d\Omega = \sin\theta d\theta d\phi$$
(3.21)

Light travels on "null intervals", with ds equal to zero in all reference systems using clocks and rulers to measure ds. The requirement that $ds = 0$ for light means the speed of light is the same in all such systems. It is also the maximum speed of information transfer. There is no action at a distance.

Event spatial separations need to be less than ct for them to be causally related. For clocks at rest, or "proper clocks", $ds = cdt_*$. For a frame where the clocks are moving with velocity v, the invariant interval is $(cdt)^2 - (vdt)^2$. Therefore, $t = t_*/\sqrt{(1-v^2)} = \gamma t_*$ and time in a frame where the clocks are moving, t, is greater than time in a frame with the clocks at rest, t_*. This is the famous "time dilation" of SR.

Moving to GR, it is postulated that matter defines the geometry of space–time. The resulting metric is no longer "flat" like Eq. (3.21) but is curved and depends on the coordinate location. Particles travel on "geodesics" which are the shortest interval between two events, like a spatial great circle route on a sphere. There are classical tests of GR which are covered in many textbooks and will be skipped here. The tests, gravitational redshift of light, the deflection of light, and the perihelion advance of Mercury are all measurements of the small effects of weak gravitational fields as measured in the twentieth century. The focus of this text is on the effects of strong gravity as explored in the present century, for example, gravity waves from black hole mergers.

The resulting GR equations relating the energy–momentum tensor (source) to the metric tensor (solution) are nonlinear, and very few closed-form solutions have been found. In this text, we explore three. The interior solution for uniform mass density was already shown in Eq. (3.12). For a solution corresponding to a classical point mass, the only parameters that are observable are mass and angular momentum. This almost exhausts all the known solutions of Einstein's GR equations.

First, there is a look at the special cases of radial and circular orbits near a Schwarzschild black hole. The metric is defined only by a non-rotating mass, M or R_s. Then general orbits are explored. Finally, general orbits near a rotating or Kerr black hole are displayed. Because the Kerr metric describes a rotating black hole, radial geodesics are impossible in any case because the space–time is dragged by the rotation so that purely radial motion is impossible.

The metric for a point mass, M, at the origin is a modified version of the SR metric with a temporal factor $(1\text{-}R_s/r)$ and a radial term modified by the inverse of that factor:

$$ds^2 = (cdt)^2(1 - R_s/r) - dr^2/(1 - R_s/r) - r^2 d\Omega^2 \qquad (3.22)$$

At large radii, the metric interval is the SR interval, as is expected. Clocks can be synchronized there and their behavior at smaller r can be tracked. For clocks at rest, $ds = cdt(1\text{-}R_s/r)$. That means clocks using coordinate time, t, go slower than clocks reading proper time, s. In the extreme case of $r = Rs$, the coordinate clocks stop and proper clocks continue to tick. When the diagonal temporal term in the metric, g_{tt}, vanishes, that behavior is called the infinite red shift point. Spatially, $ds \sim dr/(1 - R_s/r)$ and since ds is finite, dr must be zero at R_s, so that coordinate velocity, dr/dt should vanish.

Knowing the metric, the equations for geodesic motion can then be derived. In this text, they will only be quoted because they are derived in many GR textbooks. A radial geodesic will be explored first. Motion is chosen as an initial condition to start at rest from $r = r_o$. The geodesic equations for the dimensionless ratios are given as follows:

$$ds/dr = \sqrt{r/R_s}$$
$$dct/dr = \sqrt{r/R_s}/(1 - R_s/r) \qquad (3.23)$$

There is smooth behavior of s as a function of r by inspection, with $s \sim r^{3/2}$. However, ct becomes very large as r approaches R_s. These equations are solved and displayed in the App

"Geodesics_Radial_Circle". They are integrated symbolically using the Matlab symbolic utility "int".

$$s - s_o = (2R_s/3)[(r/R_s)^{3/2} - (r_o/R_s)^{3/2}]$$
$$ct - ct_o = [(r - r_o)\sqrt{r/R_s}]/(1 - R_s/r) \tag{3.24}$$

The proper time interval to fall to $r = 0$ is perfectly finite as it is classically. Coordinate time diverges at an r of R_s however, so that observers at large distances from the singularity will never see the object actually fall into the black hole. For them time stops at R_s:

$$s(0) - s_o = (2r_o/3)\sqrt{r_o/R_s} \tag{3.25}$$

The tidal force, Eq. (3.20), on an extended object leads to a tidal stress, S_T, along r which acts to elongate the traveller of length L and transverse area A. In the transverse coordinates, she is being compressed.

$$S_T \sim (mc^2 L)/(R_s^2 A) \tag{3.26}$$

For a 100-kg voyager of cubic size $1\,\mathrm{m}$, the tidal stress, with units of pressure or energy density, is about 10^{12} Pa for a black hole of solar mass when $r = R_s$. Note that the breaking stress for specialty steels is about 5×10^9 Pa, which would be disrupted to say nothing of a person. The voyager will never reach the black hole origin intact. Remote sensors are called for in order to explore these radii, made as small as possible.

The circular geodesics also have a closed-form solution. The scaled angular momentum is $(J/m)/c$ which is equal to βr classically. Also, in this and later discussions, the symbol J is used for angular momentum while L is used for luminosity. Please note also that the symbol h used here is conventional and related to angular momentum and is not to be confused with the use of h to represent the Planck constant or a metrical distortion. The radius of a classical circular orbit is given as

$$J/m = r_c v, \quad v = \sqrt{GM/r_c}$$
$$r_c = 2(J/m)^2/(R_s c^2) \sim (2h^2)/R_s \tag{3.27}$$

The expression for a circular orbit in GR in a Schwarzschild metric is, for material particles, given as

$$r_c = (h^2/R_s)[1 + \sqrt{1 - 3(R_s/h)^2}]$$
$$h = (J/mc)[1/(1 - R_s/r_c)]$$
(3.28)

The dimension of h is length. In the limit where r is much greater than R_s, the GR expression approaches the classical circular orbit.

The App "Geodesics_Radial_Circle" displays both radial and circular geodesics in a Schwarzschild space. The App figure is shown in Figure 3.17. The user chooses a starting radius for the radial geodesic motion. The equations are displayed symbolically for dr/ds and $dr/d(ct)$. The closed-form circular solution is also displayed in the figure. There are two possible plots. One is the velocity dr/ds and $dr/d(ct)$ as a function of r. At the Schwarzschild radius, the apparent

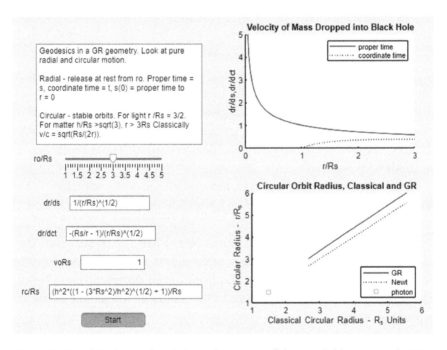

Figure 3.17:   Circular and radial geodesics in a Schwarzschild space, velocities for radial geodesics, and radii for circular geodesics.

velocity for an observer at large radius is zero, while the velocity on a free-fall trajectory has a finite velocity all the way to the origin.

The proper time and the coordinate time are plotted when that option is chosen by the user using the numerical EditField value of 2. The resulting plot is shown in Figure 3.18. The proper time is well behaved all the way to the origin while the coordinate time diverges. From the viewpoint of an observer far away, the traveller never reaches the Schwarzschild radius. Alas, for a proper observer the time is all too finite. The coordinate velocity goes to zero at R_s, while the coordinate time diverges.

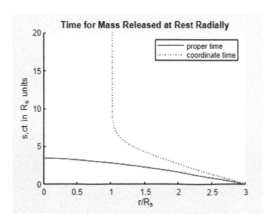

Figure 3.18: Display of the proper and coordinate time as a function of r for a radial geodesic. The proper time is well behaved while the coordinate time diverges.

For the circular geodesics, there are no stable orbits for material particles with radii less than three R_s. In GR, photons are another energy responding to the metric and subject to GR geodesic motion, as is familiar from the famous light deflection observations made during solar eclipses. Indeed, photons can have circular orbits and the orbital radius is $(3/2)R_s$.

These behaviors with strong gravitational effects are in contrast to the small solar system effects such as the perihelion advance of Mercury, or the light deflection during solar eclipses. It is gratifying that the behavior near a black hole has recently been verified. Black holes can have masses of a few solar masses. The black hole mergers that were observed as gravitational waves had masses of a few tens of solar masses. In addition, supermassive black holes with billions of solar masses located at the center of galaxies have been inferred from their effect on nearby orbiting stars. The black hole in the center of our galaxy is estimated to have roughly a million solar masses. How these immensely heavy objects were formed is still under speculation.

The first observation of the "event horizon" at R_s for a black hole was announced in April 2019. A picture of the signals from an in-falling material near a supermassive black hole was produced by a consortium of radio astronomers and is shown in Figure 3.19. Indeed, for observers using coordinate time, the material never falls in and appears near the Schwarzschild radius. This observation is a brilliant confirmation of the GR predictions when the effects of gravity are strong.

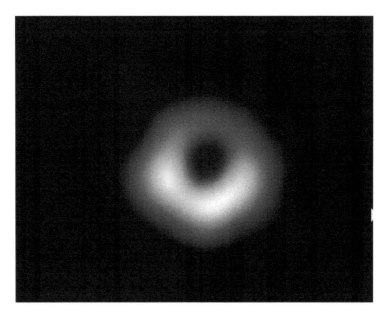

Figure 3.19: Picture of the radio emissions from a supermassive black hole at the center of a galaxy. The in-falling material is heated and radiates, but never appears to cross the event horizon at R_s.

3.10. General Geodesic Motion

The motion of light or material particles in a Schwarzschild space can be easily generalized. As in the case of classical orbits, there is a conserved angular momentum, which means the motion is confined to a plane, taken here to be equatorial, so that only the coordinates r and ϕ are relevant. The angular motion gives rise to a centrifugal potential which appears to be a repulsive force. In GR there is another added term in the equation of orbital motion, $3R_s u^2/2$. The expression in Eq. (3.29) has dimensions of inverse length:

$$d^2u/d^2\phi + u = [R_s/h^2]/2 + 3R_s u^2/2, u = 1/r \qquad (3.29)$$

This equation is like that for classical Keplerian orbits except there is an added term which acts like a classical force which is not inverse square. The GR case has the added term in general for material particles, but for light, the parameter h is infinite, which leaves only the GR term. This makes sense because it is only in GR that light, having energy, gravitates.

The orbits are displayed using the "General_Geodesic_App" and the resulting figure is displayed in Figure 3.20. The choice of orbit is made by use of a RadioButton to be Newtonian, GR for massive particle, or GR for photons. The user sets the Sliders to choose the initial position and velocity by choosing a point on the x axis, r_o, and an initial impact parameter, b. The initial velocity magnitude and direction is $du/d\phi = \cos\phi/b$ with $\sin\phi = b/r_o$. The angular momentum and number of approximate revolutions is also chosen. The user can build up an intuition by playing with the wide choice of inputs. The origin and the R_s location are also plotted on the graph.

Starting with the option for Newtonian orbits, the plot shows that the orbits repeat. That can be checked by varying the Slider for tracking in phi. The EditField J/c(circ) gives the classical J/c value for a circular orbit. Changing the Slider to ~4.7 results in a circular orbit being plotted. The specific figure shown in Figure 3.20 is for GR dynamics, option 2, near a black hole. Note that the perihelion advance is not a small effect now. A circular orbit can be obtained for

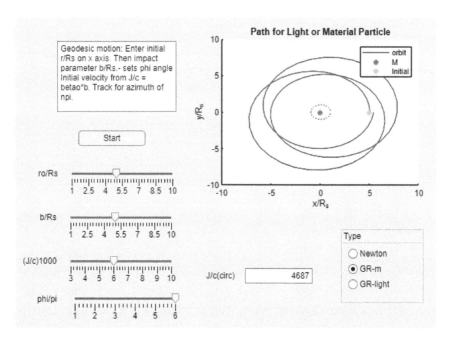

Figure 3.20:   Figure for general geodesic motion. The user chooses the initial position and velocity and the orbit is tracked for a chosen azimuthal interval. The J/c(circ) value for a circular orbit is shown numerically.

a J/c Slider value of ~5.7. For lower values, the orbit falls into the singularity, while for higher values escape is possible. If light orbits are chosen, the J/c Slider has no effect as expected since h is infinite for photons. The only variables to vary in this case are r_o and b. It becomes clear that light deflection is no longer a small effect at radii near R_s. For $r_o/R_s \sim 5$ and $b/R_s \sim 2.5$, the photon is almost captured in an orbit, for example. The user is encouraged to play with the Sliders.

3.11. Geodesic Motion in a Kerr Metric

There is a second solution for the Einstein field equations, one appropriate to a rotating black hole. Since the stars rotate, this solution is relevant. This Kerr solution is characterized by a mass, M, and a spin, J. In fact, these are the only two possible attributes of a black hole in GR theory. There is a maximum spin which occurs approximately when the azimuthal velocity classically becomes c, $J/Mc = R\beta$ and $R > J/Mc$ would mean $v < c$. The Kerr solution has a parameter a, which is a characteristic length associated with the rotation of the Kerr black hole, with a less than $R_s/2$. Centrifugal force impedes collapse and that means there is a maximum value of a for a Kerr solution.

As an aside, the Sun rotates with a period of 24 days, so that the surface velocity is $\beta = 6.9 \times 10^{-6}$. If the Sun were to shrink to a radius of $2R_s$ or $6\,\mathrm{km}$, the velocity would increase to $\beta = 0.8$.

The Kerr metric is complex and the special case of equatorial motion is chosen here. The Boyer–Lindquist formulation of the metric in spherical polar coordinates is used here. The equatorial choice maximizes the effects of rotation since at the poles they are expected to vanish. There are no radial geodesics because of "frame dragging" where the rotation pulls the space–time coordinates in the direction of rotation. Therefore, general geodesic motion must be treated.

The metric for equatorial motion is shown in Eq. (3.30). The Kerr metric becomes the Schwarzschild metric in the limit when $a = 0$. There is a non-diagonal coupling term $(cdt)(d\varphi)$, which indicates that the orbit will be "dragged" by the rotation of the star. The sign of a specifies the sense of the rotation. For example, using the SR interval and replacing ϕ by $\phi - at$ to approximate a rotating system induces a term $2a(d\phi)(dct)$ in the interval:

$$ds_K^2 = (cdt)^2(1 - R_s/r) + 2a(R_s/r)(cdt)(d\phi) - (dr)^2 r^2/\Delta$$
$$- [r^2 + a^2(1 - (R_s/r)](d\phi)^2$$
$$a = J/(Mc), \quad \Delta = r(r - R_s) + a^2 \tag{3.30}$$

There is a boundary of infinite red shift where the diagonal time factor in the metric vanishes which means coordinate time diverges as in the Schwarzschild metric case, at $r = R_s$. There is an equatorial event horizon where the radial inverse metric coefficient vanishes as it does for the Schwarzschild metric but not at r of R_s but rather when $\Delta = 0$. The region between the inner and outer boundaries is called the ergosphere. The inner boundary is the event horizon while the outer boundary is the static limit, or infinite red shift boundary, where a particle can remain at a fixed azimuth. Between the two limits a particle must be dragged in the direction of the rotation of the black hole. The two boundaries overlap at the rotational axis as expected intuitively. On the equatorial axis,

$$r_{\text{red}} = R_s$$
$$r_{\text{eh}} = R_s[1 + \sqrt{1 - (2a/R_s)^2}]/2 \qquad (3.31)$$
$$a < R_s/2$$

Orbits, in general are characterized by an energy parameter, U, and an angular parameter, J. This is similar to the case of Keplerian orbits which have two constants of motion, energy and angular momentum. Classical orbits will be explored in detail in the section immediately following in the text.

Orbits in a Kerr space are explored using the App "Kerr_Orbits_2". The user chooses parameters corresponding to the classical energy, U and the angular momentum J, and the starting value of the radius using the Sliders that are supplied. There is also a choice of ingoing or outgoing orbits.

The parameter a is fixed at 0.98 in order to maximize the influence of the rotation and the initial values of s, t, and φ are zero. The path of the test particle is plotted with markers for equal intervals of coordinate time in order to give an impression of the velocity seen by observers at large radii at each point on the orbit. This is a "movie" in terms of coordinate time.

The geodesic equations for $ds/d(ct)$ and $d\varphi/d(ct)$ are given as follows:

$$ds/d(ct) = Q = \Delta/[U R_o^2 - (R_s a J)/r]$$
$$d\phi/d(ct) = \omega = [JQ + (R_s a)/r]/R_o^2 \qquad (3.32)$$
$$\Delta = r^2 + a^2 - R_s r, \quad R_o^2 = r^2 + a^2 + (R_s a^2)/r$$

The equation for r is second order and the Matlab utility "ode45" is used to solve it by treating it as two first, order equations. The actual radial equations are complex and can be viewed in the App if desired, but they are not very edifying.

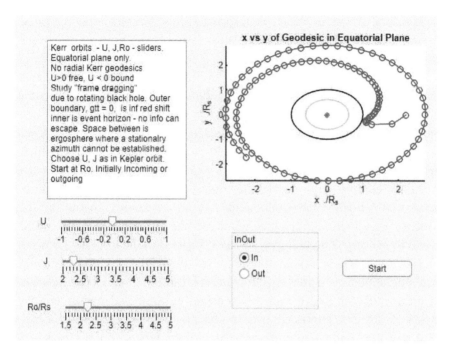

Figure 3.21: Figure generated by the App "Kerr_Orbits". Motion is only in the equatorial plane.

The figure created by the App is shown in Figure 3.21. Note the "frame dragging" where the initial azimuthal velocity is reversed by the effect of the rotation of the Kerr black hole. Since most stars rotate, the effects on orbits near singularities may need to be taken into account. In fact, energy can be extracted from the ergosphere by removing some of the energy of Kerr rotation, at least in principle. The extracted energy is matched by a reduction in the rotation of the Kerr black hole.

The user can vary the inputs to create a variety of orbits. Orbits that look like classical hyperbolic orbits occur with positive U values and well-chosen J and R_o values. For example, $U = 1$, $J = 4$ and $R_o/R_s = 2$ looks like a classical orbit. For bound cases with $U = -1$, the orbit is dragged to the ergosphere event horizon. For the particular case shown in Figure 3.20, the object closely approached the outer ergosphere boundary, slows down (movie), then speeds up and enters an orbit which is approximately circular, having been dragged into a counterclockwise direction by the spin of the Kerr singularity.

There is a weak field metric solution for rotation which is the Lense–Thirring metric. The radial and temporal factors are $(1 + R_s/r)$ and $(1 - R_s/r)$ respectively. There is a cross term, LT, which multiplies $d\phi d(ct)$, which is $2aR_s/r$, $a = J/(Mc)$ in the Kerr metric, Eq. (3.30). For the weak field, $LT = 2(R_s/r)(J/Mc)$, which agrees with the Kerr cross term.

This chapter ends with the singular cases for the endpoints of stars. It began with a look at how stars might cluster out of protons and helium nuclei which were the products of primordial nucleosynthesis. The collapse of these protostars heated the protons to the point where fusion might start. Typically, a main sequence star might be in a stable equilibrium for a billion years. The interior structure of the stars can be reasonably modeled using simple classical physics. However, eventually, any star will run out of fuel that can be fused. The endpoints for a star depend on the mass. If collapse is stabilized by electrons, a white dwarf is formed. If only neutrons can stabilize the mass, a neutron star will form and may be observed as a pulsar. If even the neutrons cannot stop the collapse, a black hole can form, with a mass and perhaps a spin.

There is much information to be gained from ground-based astronomy and space-based observatories for all portions of the electromagnetic spectrum, and just recently, the gravitational wave spectrum has become available. Nevertheless, it is of great interest to speculate on the possibilities of humans traveling to all parts of the solar system, and perhaps even to nearby stars. Those possibilities are the subject of Chapters 4 and 5 of this book.

Chapter 4

Apps for Solar Exploration

"And yet it moves."

— **Galileo Galilei**

"It is not easy to see how the more extreme forms of nationalism can long survive when men have seen the Earth in its true perspective as a single small globe against the stars."

— **Arthur C. Clarke**

"Now voyager sail thou forth to seek and find."

— **Walt Whitman**

4.1. Eccentric Kepler

In Chapter 2, the formation and subsequent steady equilibrium of stars was explored. Then in Chapter 3, the endpoints of stellar evolution were examined; white dwarfs, neutron stars, pulsars, and black holes (static and rotating). In this chapter, the focus shifts to the solar system and the ways it has been and may be explored. As it is said, "small steps". Finally, in Chapter 5 possible visits to other stellar systems and exoplanets are discussed.

The first step in solar exploration is to escape the gravity well of the Earth. The orbits around a point mass for an inverse square force are ellipses if the orbit is bound. They are re-entrant, which means the perihelion, or point of closest approach to the Sun, occurs at the same point in space on every orbit. We have already seen that GR does not have re-entrant orbits, but the approximation is good for the solar system and Newtonian physics will be used exclusively in this chapter.

The purely geometric aspects of ellipses are the semi-major and semi-minor axes, a and b and the two distinct foci located at $x = ae$ and $x = -ae$. The eccentricity, e, indicates the difference between a and b. With distance r_a and r_b from the foci, a point on the ellipse is the locus where $r_a + r_b = 2a$. The radius of the perihelion is a $(1 - e)$ and the aphelion, point of furthest approach, is $a(1 + e)$. The area of the ellipse is $A = \pi ab$.

$$(x/a)^2 + (y/b)^2 = 1, \quad x = a\cos\phi, \quad y = b\sin\phi$$
$$\text{foci } (\pm\sqrt{a^2 - b^2}, 0), \quad e = \sqrt{1 - (b/a)^2} \qquad (4.1)$$
$$r = a(1 - e^2)/(1 + e\cos\phi)$$

The orbits are in a plane for a central force which has a constant of the motion, J, the angular momentum. The equatorial plane (r, ϕ) will be assumed in this text. The gravitational potential, Φ, and the potential energy, V, are

$$\Phi = -GM/r = -R_s c^2/(2r), \quad V = \Phi m, \quad V/mc^2 = -(R_s/2r)$$
$$(4.2)$$

Since the Earth is at a distance of one astronomical unit (AU) or 1.5×10^{11} m and the solar Schwarzschild radius is 3×10^3 m, the gravitational potential energy is very weak on the scale of the rest mass thus justifying a classical approach to solar system physics.

The total energy, U, for a bound state, $V + K$, is negative, indicating that it will take energy to escape from a bound state. The virial theorem connects the average value of the potential and kinetic energies, $K, 2\langle K \rangle + \langle V \rangle = 0$, i.e.,

$$\langle V_G \rangle = -2\langle K \rangle, \quad \langle U \rangle = \langle V_G \rangle/2 \qquad (4.3)$$

Because of the conservation of angular momentum, an effective centrifugal potential can be defined from which the centrifugal force, F_c, is derived by differentiating with respect to r, where ω is the circular frequency for a circular orbit. The potential is an effective repulsive inverse square potential corresponding to an inverse cube force.

$$F_c/m = \omega^2 r, \quad \Phi_c = (\omega r)^2/2 = (J/m)^2/(2r^2) \qquad (4.4)$$

The total energy, U, depends only on the semi-major axis, a, and the orbiting mass. The semi-minor axis, b, depends on the angular momentum in addition to U. The orbital period, τ, depends on the three halves power of the semi-major axis which is one of the laws of Kepler regarding planetary motion.

$$U = -(GMm)/2a, \quad b = J/\sqrt{2m|U|},$$
$$\tau = 2\pi a^{3/2}/\sqrt{GM} \qquad (4.5)$$

Starting from an initial radius on the orbit, r_o, at the perihelion, which is just the radius for a circular orbit of zero eccentricity, the geometry of the orbit is

$$r = r_o(1+e)/[1+e\cos(\phi)] = a(1-e^2)/[1+e\cos(\phi)] \qquad (4.6)$$

In order to illustrate the orbits geometrically, a basic Matlab .m file, "Kepler_eccentric_2", is invoked because the algebra is very simple. The parameter r_o is r at zero azimuthal angle. In the script, r_o is chosen to be equal to 1. The plotted orbits are for several eccentricities, from 0, a circle, to 0.96, an eccentric ellipse, to 1.2, an unbound hyperbola. The results are plotted in Figure 4.1. The most deeply bound orbit is circular. The orbits for larger eccentricities become less bound and the ellipses have smaller b/a ratios. The Sun is at a focus, placed at the origin since, by assumption. The Sun is at the Center of Momentum (CM). Finally, a bound orbit to free orbit transition occurs for a parabola, with e value of 1. Larger values of e correspond to unbound or hyperbolic orbits.

These orbits have removed any information provided by time, such as the velocity on the orbit. That reflects the fact that for an inverse square attractive force, the orbits repeat over time. In order to explore the time development of the orbit, the differential equations of motion can be solved explicitly. This is an alternative way of looking at the orbits. In some applications, time is useful, but in others, energy is more useful. In this text, the chosen treatment will be application specific.

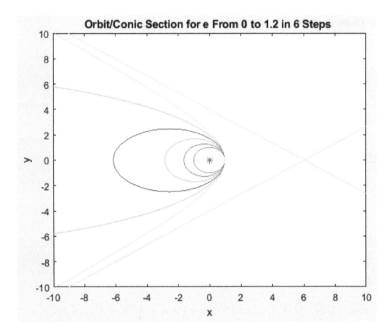

Figure 4.1:   Plots of the orbits of both bound and unbound systems with e going from 0 to 1.2. The orbit for $e = 1$ is parabolic. The red star represents the Sun at the origin. The circular orbit has a radius of one unit.

4.2. Time Orbits

A straightforward approach is to solve the equation of motion numerically using the Matlab utility "ode45". Doing so allows the user to see a "movie" made in "frames" of equal time which then gives a sense of the speed on all parts of the orbit. In this case, motion is in the plane (x, y) and a central inverse square force is assumed. The mass M is assumed to be much larger than the planetary mass m, so that M is fixed in space. Later, this assumption will be relaxed, but for now it is a useful approximation in the case of the solar system:

$$d^2\vec{r}/dt^2 = -GM(\vec{r}/r^3) \tag{4.7}$$

The user supplies the initial conditions for x and y positions and velocity and the numerical solution are displayed for x and y in equal time steps. The App "Kepler_Orbit_Time" allows the user to look at the time development of both bound and unbound orbits. Sliders allow the user to choose the initial position $(r_o, 0)$ and the initial velocities v_r and v_ϕ. The units of distance are AU. The units for time are years. One such orbit is shown in Figure 4.2. The orbit is bound, eccentric, and the velocity is large when it is nearest to the force center. The user can generate many interesting orbits, bound and unbound, using the Sliders. As always, play is a useful way to build up intuition.

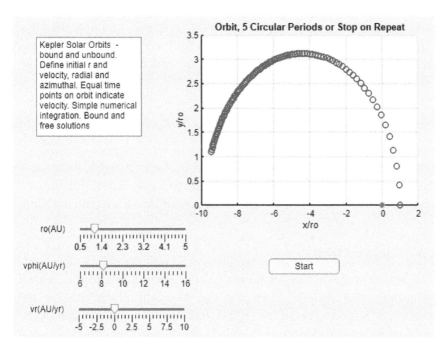

Figure 4.2:   A "movie" of a specific orbit chosen using the three Sliders. The red star indicates the Sun as force center. The plotted points are evaluated in equal time intervals. The velocity is largest nearest the Sun. The eccentricity of the elliptical orbit is large.

It can be asked "why not integrate the equations of motion"? That integration has two constants of motion, total energy and angular momentum. In Eq. (4.8) the kinetic energy, K, in the equatorial orbital plane is shown. The azimuthal contribution to K can be removed by using the angular momentum. That then leads to an effective potential, as in Eq. (4.4), which is purely radial and has terms due to gravitational attraction and centrifugal repulsion. The problem has been reduced to one dimension:

$$K = (m/2)[(dr/dt)^2 + r^2(d\phi/dt)^2], \quad J = mr^2(d\phi/dt)$$
$$U = K + V = (m/2)(dr/dt)^2 + [V + J^2/(2mr^2)] \quad (4.8)$$

However, the Euler–Lagrange equation for r does not lead to closed-form solutions but rather elliptic integrals. One standard approach to that problem is a change of variable to $u = 1/r$. Then the time dependence can be removed in order to find the differential equation for u as a function of ϕ. This form was quoted in Eq. (3.29) with an added GR piece:

$$d^2u/d\phi^2 + u = [GM/(J/m)^2] = [R_s c^2/2(J/m)^2] \quad (4.9)$$

In what follows, sometimes the energy variables are used, sometimes the time and velocity variables are featured. The choice depends on what approach seems to offer the most insight into a specific problem.

4.3. Energy Orbits

The brute force solution of numerical integration of the differential equations using "ode45" does not use some simplifications for the problem. The conservation of U and J allows for the reduction of the problem to a 1D differential radial equation, with an effective radial potential which accommodates the centrifugal force, Eq. (4.9). The crucial parameter is the energy U scaled by the energy for a circular orbit, or q. The relevant energy equations appear in Eq. (4.5).

For the energy ratio less than -1, there is no stable orbit and the mass crashes into the Sun. For a value of $q = -1$, a circular orbit exists. A circular orbit is the most deeply bound possible orbit. For energy greater than zero, the orbit is unbound, hyperbolic, or parabolic. For intermediate energies, the orbit is less bound than the spherical orbit and is elliptical. The turning points, or points on the semi-major axis of the ellipse where the x velocity is zero, are displayed along with the major and minor axes of the ellipse in Eq. (4.10):

$$q = U/(GMm/2r_o), \quad e = \sqrt{1+q}$$
$$a = 1/|q|, \quad b = a\sqrt{1-e^2} \tag{4.10}$$
$$x_{\text{turn}} = (1/q)(-1 \pm e)$$

The App "Kepler_Orbit_Energy" displays orbits as functions of energy, U. The potential is the effective potential shown in Eq. (4.8). This potential will appear again later in the discussion of Lagrange points. Defining a parameter q with r_o fixed at 1 AU, the eccentricity is e, the semi-major and semi-minor axes are a and b and the turning points for the ellipse are x_{turn}.

The figure created by the App is shown in Figure 4.3. The value of q is chosen by the user using a Slider. Since the r_o value

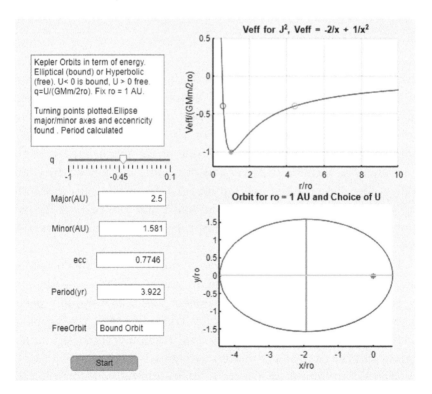

Figure 4.3:   Energetics of solar orbits. The r_o value is fixed at 1 AU. The Slider sets the energy which defines the ellipse axes, the eccentricity, the orbital period, and the turning points.

is fixed, the derived values for a, b, e, and τ are calculated and displayed in EditFields. The turning points are shown on the first plot of the effective potential. The blue star is the minimum value of the potential, that for a circular orbit. The second plot shows the elliptical orbit, if bound, with the major axis in green, the minor axis in red and the ellipse itself in blue. The focus is the location of the Sun, at the origin.

Using the Slider to set $q = -1$, the user sees a circular orbit, $a = b, e = 0$, period = 1 year, and both turning points at r_o. The plots shown in Figure 4.3 are for an eccentric but bound orbit. The turning points in the upper plot for the effective potential can be identified in the lower plot for the ellipse. Using the Slider, positive energy solutions can also be generated and an EditField (Text) indicates that the orbit has become free.

4.4. Perihelion Advance

Early in the twentieth century, the perihelion advance of Mercury was a major problem in astronomy. In Newtonian mechanics, an inverse square force predicts perfectly re-entrant orbits. The planet Mercury is observed to advance the perihelion of its orbit by about 5,600 s of arc per century. After corrections due to the perturbations of the other planets, a residual precession of about 43 s per century was found. Indeed, so far the assumptions for the orbits have been the simplest; that the Sun was fixed in space and the planets did not influence one another. These assumptions will be explored further later in the text.

The invisible planet Vulcan was hypothesized to perturb Mercury. Alternatively, the inverse square law of Newton was thought to perhaps not have exactly the exponent 2. These were ugly paste, up fixes that were only motivated by trying to "fix" the problem. Einstein solved the problem in an elegant fashion without any ad hoc kludges and the equation of motion was already quoted in Eq. (3.29), with the GR addition of a term in the potential $\sim R_s/r^3$. Solving that equation or, more easily, treating the added GR term as a small perturbation, the advance of the perihelion of Mercury per orbit is

$$\delta\phi \simeq 6\pi GM/[ac^2(1-e)^2] \simeq 3\pi(R_s/a)_{e=0} \qquad (4.11)$$

In Eq. (4.11) a is the semi-major axis of Mercury and e is the eccentricity. Numerically, the phase advance is about "0.1" per revolution. The cumulative advance is "43" per century which agrees almost perfectly with observation. Clearly, Mercury, having the smallest planetary value of a, is the planet most likely to have a measurable phase advance. This GR prediction was a strong early indication that GR was a correct theory of nature, although the GR effect is small, to say the least.

To illustrate the point, a simple numerical integration using "ode45" is used to look at a largely inverse square force law with a small additional inverse fourth-power force in the App, "Perih_App". A Slider allows the user to set b and, in particular, check that $b = 0$ on the Slider yields re-entrant orbits. The main force has units of GM fixed at 1. The treatment is perturbative in order to keep the calculation simple. As seen in Figure 4.4, which is a case of a strong

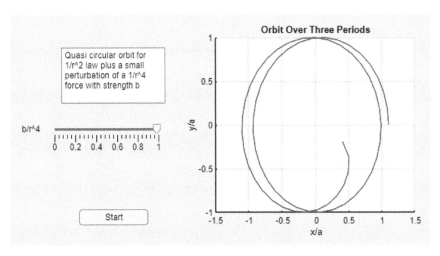

Figure 4.4: Plot of the orbit of a particle under the influence of a dominant inverse square law and an inverse fourth-power force which is quite strong in this particular case.

additional force in order to make the changes evident, the orbits do not close and advance the perihelion location.

4.5. Simple Rocket — Symbolic

Planetary exploration is 50 years old. Almost all of the travel has used the principle of the rocket. Rocket motion is long known, having been pioneered by the Chinese. Rocket motion is described using the conservation of momentum. A rocket, of mass m, is powered by ejecting a mass, dm with a velocity, v_o, thus accelerating the remaining rocket mass and gaining velocity dv. The exhaust velocity is v_o and the total time of fuel consumption is T assuming that there is no payload and that the rocket engines and plumbing are also weightless. The fuel burning rate, dm/dt, is assumed to be constant. No other force acts on the rocket at this simplest level of description. The acceleration of the rocket is v_o/T initially and increases as fuel is burned up and ejected, which lightens the rocket.

$$
\begin{aligned}
mdv &= -v_o dm \\
T &= m_o/(dm/dt) \\
m &= m_o(1 - t/T) \\
d^2y/dt^2 &= dv/dt = v_o/(T - t)
\end{aligned}
\tag{4.12}
$$

These equations are solved in the case where there is a payload in the App, "Rocket_Sym_Pay", using the symbolic differential equation solver "dsolve" applied to the last line of Eq. (4.12). In this case, the payload mass is m_p and the burn time is reduced since the final mass is no longer the idealized value of zero. By payload what is meant is the mass of the rocket after a burn time of t_p. The final velocity is the exhaust velocity times the logarithm of the ratio of the payload mass to the initial mass:

$$
\begin{aligned}
m_p/m_o &= (1 - t_p/T) \\
v(m_p) &= v_o \ln(m_p/m_o)
\end{aligned}
\tag{4.13}
$$

The maximum velocity of the rocket is defined by the payload mass. The main issue with rockets is that the final velocity depends only weakly, logarithmically, on the payload ratio, but linearly on the exhaust velocity. Clearly, fuels with the largest exhaust velocity are favored if cost and safety are not the determining factors.

For the App, the user chooses the fractional payload mass m_p/m_o with a Slider. The solutions for $v(t)$ and $y(t)$ are calculated symbolically and displayed in the App figure using EditFields. The plots are for $v(t)$ and $y(t)$ and t is only plotted to a maximum value of t_p/T. The plots are made in equal time "movie" frames so that the relative changes at different times are displayed. The velocity change is illustrated in the top plot. In the display of the symbolic solutions, Matlab uses "log" for the natural logarithm, base e, and "log10", for base 10 (Figure 4.5).

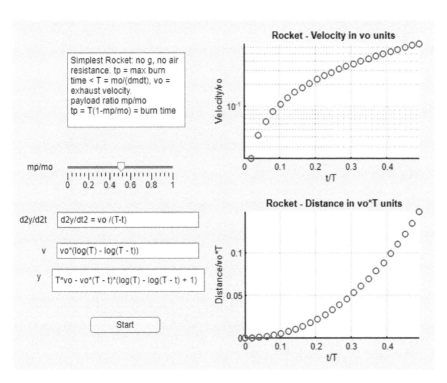

Figure 4.5: Plots of velocity and position, $y =$ height, as a function of time for a rocket with a user-chosen payload ratio. Plots only extend to a maximum time of t_p.

With a payload, the maximum, but finite, acceleration occurs at the end of the burn:

$$a_{\max} = (v_o/T)(m_o/m_p) \tag{4.14}$$

There is a clear advantage to use fuel with the highest possible exhaust velocity since the final velocity depends linearly on v_o. Some possible options are provided in the table shown in Figure 4.6. Liquid fuels are favored over most solid fuels, at least from the point of having a large v_o. Clearly, liquid hydrogen is favored over kerosene ($\sim$3 km/s) used in the Saturn rocket first stage. However, the liquid requires cryogenic support systems and has safety issues. Because of this, use of liquid hydrogen is typically confined to the upper stages of multi-stage rockets. In order to exceed the exhaust velocity, a small payload to total rocket mass is required. In the references, "specific impulse" may be mentioned. It is the momentum change per mass of propellant. For rockets, it is simply the exhaust velocity. A rocket needs to carry both an oxidizer and a fuel to burn.

ROCKET PROPELLANT PERFORMANCE				
Combustion chamber pressure, P_c = 68 atm (1000 PSI) ... Nozzle exit pressure, P_e = 1 atm				
Oxidizer	Fuel	Hypergolic	Mixture Ratio	Specific Impulse (s, sea level)
Liquid Oxygen	Liquid Hydrogen	No	5.00	381
	Liquid Methane	No	2.77	299
	Ethanol + 25% water	No	1.29	269
	Kerosene	No	2.29	289
	Hydrazine	No	0.74	303
	MMH	No	1.15	300
	UDMH	No	1.38	297
	50-50	No	1.06	300
Liquid Fluorine	Liquid Hydrogen	Yes	6.00	400
	Hydrazine	Yes	1.82	338
FLOX-70	Kerosene	Yes	3.80	320
Nitrogen Tetroxide	Kerosene	No	3.53	267
	Hydrazine	Yes	1.08	286
	MMH	Yes	1.73	280
	UDMH	Yes	2.10	277
	50-50	Yes	1.59	280
Red-Fuming Nitric Acid (14% N_2O_4)	Kerosene	No	4.42	256
	Hydrazine	Yes	1.28	276
	MMH	Yes	2.13	269
	UDMH	Yes	2.60	266
	50-50	Yes	1.94	270

Figure 4.6: Some typical liquid fuels for rockets. The specific impulse is a parameter closely related to exhaust velocity. Liquid hydrogen is the best fuel and is often used with a liquid oxygen oxidizer. Rockets need both a fuel and an oxidizer, while jets can use the Earth's atmosphere as an oxidizer.

In order to start exploration of the solar system, the first, small, step is to boost out of the gravity well of the Earth and go into low Earth orbits (LEO). This feat has only recently been achieved, in 1957, with the Russian "Sputnik" satellite. An LEO is a few hundred kilometers above the surface of the Earth and has an orbital velocity of about 8 km/s. A glance at Figure 4.6 shows that quite small payload ratios of final to initial mass will be needed even for this first step.

4.6. Rocket on Earth — Saturn 5

The NASA Saturn 5 was the rocket used to go to the Moon and return. In order to put a small payload into LEO, a final velocity of 7.9 km/s is needed. In order to escape the gravity well of the Earth or the Sun, a larger final velocity is needed, Eq. (4.15), where M refers to the mass of the Earth or the Sun and r is the distance away from the attractive mass. The escape velocity for the Earth is 11.2 km/s while to escape the Sun starting from 1 AU away takes 42.1 km/s. By the way, launch sites are often located near the equator so that the rocket can use the orbital velocity of the Earth, $v = \omega r$, as an assist, gaining 0.46 km/s at the equator:

$$v_{esc} = \sqrt{2GM/r} \qquad (4.15)$$

The Saturn rocket was first launched in 1967 and continued until 1973. It is a three-stage rocket. A toy model is made here of a single-stage rocket. Saturn had a LEO payload of 1.2×10^5 kg and a launch mass of 2.9×10^6 kg or a 4.1% payload ratio. The exhaust velocity was 2.6 km/s with a burn time of 170 s.

In the App "Saturn_Escape_App", the tradeoff of final velocity and payload ratio can be easily explored. In this case, there is another term in the acceleration equation. The term is assumed to be constant because it is applied to LEO, $g = 9.8$ m/s^2 (Figure 4.7):

$$g = GM/R_E^2 \qquad (4.16)$$

This attraction slows down the rocket and limits the payload ratio. In the App, a sufficient acceleration is assured for lift off which limits

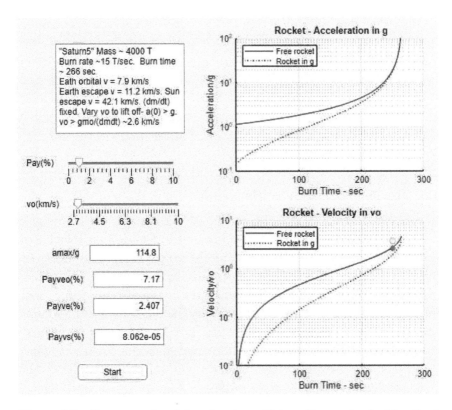

Figure 4.7: Figure produced by the App "Saturn_Escape_App". The fixed parameters are approximately those of the Saturn 5 rocket, but the exhaust velocity is somewhat larger. The payload mass percentage can be chosen by the user. The maximum acceleration occurs at the end of the burn but is not an issue because of the short time duration.

the exhaust velocities to be such that $(v_o/m_o)(dm/dt) > g$ at launch time, $t = 0$. The velocity is taken at lift off to be $v(0) = 0$. This is not really the multi-stage historical Saturn 5 but a simplified model of it. The total burn time, T, is reduced since the burn stops when the remaining mass is the "payload" mass. The burn time, acceleration, and velocity taking the gravity of the Earth, into account are given as follows:

$$m_p = m_o - (dm/dt)t_p$$
$$dv/dt = (v_o)(dm/dt)/[m_o - (dm/dt)t] - g \qquad (4.17)$$
$$v(t) = -v_o \ln[1 - (dm/dt)t/m_o] - gt$$

The App figure that is generated is shown in Figure 4.7. The rocket acceleration and velocity as a function of the time are plotted for both a free rocket and a rocket in the Earth's field. The rocket total mass and dm/dt are fixed. The payload as a percent of the initial mass is chosen by the Slider. The exhaust velocity can also be chosen by the Slider. On the plot, the red dot shows the LEO velocity which is needed, while the green circle shows the Earth's escape velocity. With the chosen Slider values, "Saturn 5" can attain LEO even when the motion taking g into account is considered. The rocket in the Earth's field, with g, may start with almost no acceleration. The maximum acceleration, in g units, is displayed for the end of the burn.

However, compared to the free rocket, the rocket in "g" largely catches up near the end of the burn time. The payload percentages to attain Earth's orbit, Earth's escape, and solar escape are displayed in the free-rocket approximation, Eq. (4.18), but the user should tune the payload mass and v_o Sliders to find a set that just attains LEO. The low percentage for the Sun means that a simple approach to exiting the solar system is not going to work well as will be explored later. For now, "small steps".

Ignoring the effect of g, the payload percentage needed to acquire a final velocity v_{esc} with an exhaust velocity v_o is given as

$$m_p/m_o \sim e^{-v_{\text{esc}}/v_o} \tag{4.18}$$

A photo of the actual Saturn 5 near liftoff is shown in Figure 4.8. In fact, the rocket has a multi-stage configuration. This is advantageous in achieving large final velocities as will be explored later in the text. The Command Module was 0.006 of the total mass at launch and the exhaust velocity of the main stage was 2.6 km/s.

4.7. Rocket with Air Drag

So far the effects of the atmosphere have been ignored. However, they are not negligible because the atmosphere exerts a velocity-dependent drag force. In addition, it is useful to study the effects of the atmosphere because in the future rockets may land on planets with different atmospheres than the Earth. In any case, returning to

Figure 4.8:　Saturn 5 rocket at liftoff. The upper part of the rocket contains the second and third stages.

Earth will mean exposing the rocket to drag forces as it sheds its initial velocity.

The App "Saturn_Escape_Drag" looks at the further reduction in payload fraction due to the drag of air resistance during the launch. Acceleration now has a component of variable "g" with altitude in order to be a bit more realistic and a variable air drag because the velocity changes during the burn as well as the change of atmospheric density with altitude. Drag scales as the square of velocity and the atmospheric density falls off approximately exponentially with height with a scale $y_{\mathrm{atm}} = 9.1\,\mathrm{km}$ (Figure 4.9):

$$d^2y/dt^2 = v_o(dm/dt)/m - g[R_E/(R_E + y)]$$
$$- [(\mathrm{visc})v^2 e^{-y/y_{\mathrm{atm}}}]/m \qquad (4.19)$$

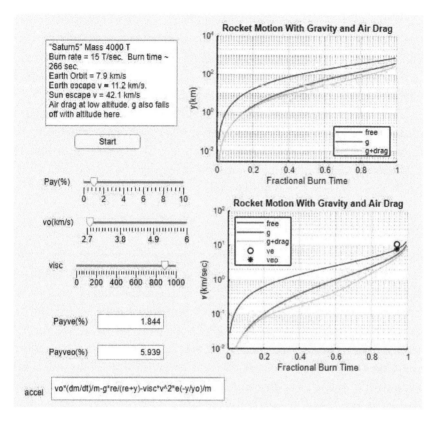

Figure 4.9:   Results for a rocket with g and drag reductions in the pure rocket acceleration. The rocket with the Slider values shown goes into LEO at a height of about $200\,\mathrm{km}$ with a payload of $\sim 1\%$ or $4 \times 10^4\,\mathrm{kg}$.

The values of m_o and dm/dt are again fixed in this example. The exhaust velocity is chosen via the Slider setting as is the payload mass percentage. The viscosity is also chosen over a wide range by Slider. The expanded range allows the user to obtain some intuition as to how drag forces affect the trajectory. The payload fractions are quoted for LEO and Earth's escape velocity in the free rocket approximation, as before. The equation of motion is solved numerically with "ode45" and the more correct payload fractions can be found by varying the Sliders.

The free-rocket payload fraction is about 6%, while the more correct fraction is about 2% for the results shown. The shape of the trajectory of the rocket with drag is very characteristic. At early times, the velocity is low and so the drag effect is small. As the velocity builds up, the effect of drag is more noticeable. However, at even higher altitudes, the atmosphere falls off and the drag effect therefore falls off again due to reduced atmospheric density.

Having attained the orbit, the passengers will later want to return to Earth. The re-entry is an unpowered inverse process which will be explored later in the text. If the payload is 1 T or 1,000 kg, it has an LEO velocity of about 8 km/s or a kinetic energy of about 3.2×10^{10} J, 32 GJ, which needs to be dissipated.

4.8. Two-stage Rocket

The small payload issue is daunting. However, it is clear that after burning the fuel the payload is the actual desired mass plus the dead weight of plumbing needed to efficiently burn the fuel. One way to partially evade the problem is to have multiple stages. A first stage burns all its fuel and then disengages so that its dead weight no longer needs to be carried. A second stage, with a similar fractional weight of plumbing and payload, then fires. Indeed, many missions use multiple stages, for example, the space shuttle has solid-state booster rockets that are dropped off after they are exhausted. The Saturn was a three-stage rocket assembly.

For the rockets, general results for initial and final velocities in terms of initial and final masses are given as follows:

$$v_f = v_i + v_o \ln(m_f/m_i)$$
$$m_f = m_i - (dm/dt)(t_f - t_i)$$

(4.20)

The payload as a fraction of the total rocket mass is fixed for a single-stage and two-stage configuration in the App "Saturn_2Stage_App". The total mass is fixed as is the constant burn rate. The plumbing mass ratio to the fuel, R, is the same for both stages as is the exhaust velocity in this simplified model. The final velocity in units of v_o for the one-stage and two-stage options is shown in the App plot. The velocity as a function of time is also plotted. The burn time is reduced from previous discussions because there is now reduced fuel since some of the weight is dead plumbing in both stages (Figure 4.10).

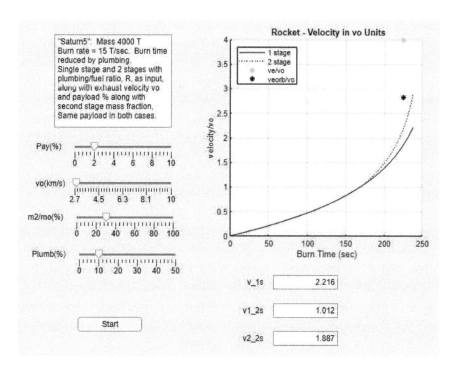

Figure 4.10:   Results comparing the velocities of a single-stage and two-stage configuration with the same total weight and true payload weight. The numerical values displayed are in v_o units.

The fuel mass is the difference between initial and final mass, corrected for the plumbing using the parameter R. The fuel mass burns for a time defined by dm/dt and the final velocity is that of the payload plus upper stage plumbing at the end of the fuel consumption. The single-stage rocket's final velocity is v_{f1}, while the final velocity of the second stage is v_{f2}:

$$m_{fu} = (m_i - m_f)/(1 + R), \quad t_{fu} = m_{fu}/(dm/dt), \quad R = m_{pl}/m_{fu}$$
$$v_f = v_i - v_o \ln[1 - (dm/dt)t_{fu}/m_i]$$
$$v_{f1} = -v_o \ln[1 - (1 - m_p/m_o)/(1 + R)] \qquad (4.21)$$
$$v_{f2} = -v_o\{\ln[1 - (1 - m_2/m_o)/(1 + R)]$$
$$- \ln[1 - (1 - m_p/m_2)/(1 + R)]\}$$

The results of the App are shown in Figure 4.10. The Sliders are used to pick the payload mass with respect to the full initial mass. The exhaust velocity can be changed as can the ratio of the full second-stage mass to the initial mass. Finally, the ratio, R, of plumbing to fuel mass is a variable that can be set. For the configuration defined in the App, there is a maximum two-stage final velocity with the second-stage mass close to 30% of the total mass. The useful payload is about 2% and the plumbing to fuel ratio is about 10% for all stages.

For this particular choice of parameters, the single stage attains only 76% of the final velocity in the two-stage case. The user is encouraged to play with the Sliders and see how things scale. It appears that a two-stage rocket with these simplified parameters could put a payload of 80,000 kg into an LEO. Clearly, there is a great advantage in using multiple-stage rockets since there is no reason, aside from complexity, to drag around plumbing dead weight after the fuel is consumed. Indeed, most rocket launches use multiple stages for that reason. In fact, to reduce costs, the first stage has recently been shown to be able to make a soft landing so that it can be reused. This feature significantly reduces costs. In the past the first stage of a mission was simply discarded.

The Saturn 5 first stage had a total weight of $\sim 2.3 \times 10^6$ kg. The second stage weighed 5×10^5 kg or 22% of the first stage. Properly configured, it put a payload of 1.4×10^5 kg into LEO which remains a record to the present day.

4.9. Docking

The focus for rockets has been, so far, simply to escape the gravity well of the Earth. Having achieved LEO with a two-stage rocket, the discussion now turns to docking, followed by landing on the Moon and then returning to Earth. The aim is to dock with the International Space Station (ISS). One possible way is to go a bit faster than the ISS. The strategy in this case is at every moment to aim at the present location of the ISS by making course corrections. When contact is made, a breaking must be done, of course. The rocket is steered and is no longer in ballistic motion.

This strategy can be described by a differential equation with a closed-form solution. The equation for the altitude $y(t)$ value for the rocket is given as

$$d^2y/dx^2 = \sqrt{1 + (dy/dx)^2}/[q(L - x)] \qquad (4.22)$$

The rocket location is $x(t), y(t)$ while the ISS is approximated to be at an altitude, y of L, and to move in x with a fixed horizontal orbital velocity. The parameter q is the velocity ratio of the rocket upper stage and the ISS. As q approaches 1, the horizontal distance diverges.

The solution to this differential equation is found in the App "Docking_Intercept". The equation is solved symbolically, in closed form, using the Matlab utility "dsolve". The default variables in "dsolve" are defined to be t and x, which the user needs to remember. The user picks a value for L and a value for the ratio of speeds, q, of the second stage and the ISS using the Sliders. The intersection point at y of L is shown in an EditField as x_{int}. The figure generated by the app is displayed in Figure 4.11 for a particular set of Slider choices. A "movie" is made of the paths of the ISS and the rocket. A second plot shows how much the rocket has to "lead" the ISS to interception as a function of the speed ratio. A small value of q takes a longer time to dock but makes the breaking needed at the docking less of an issue. Other docking strategies can be imagined; this is only a simple and solvable one.

4.10. Lunar Lander

At some point, the rocket passengers will want to return to a solid body. The easiest one is the Moon which was first accomplished 50 years ago. There is no atmosphere to worry about, and the gravity well is not as deep as that of the Earth. At one point, also about 50 years ago, the problem of how to land on the Moon was a video game called "lunar lander". The user was given a rocket with a finite amount of fuel with a specific exhaust velocity. In this case, an analytic solution is used to optimize the solution.

Assume the lander starts at rest at an altitude y_o above the surface of the Moon. The lander goes into free fall for an optimal time t_{ff} and then fires the landing rocket at t_r with an acceleration a, such

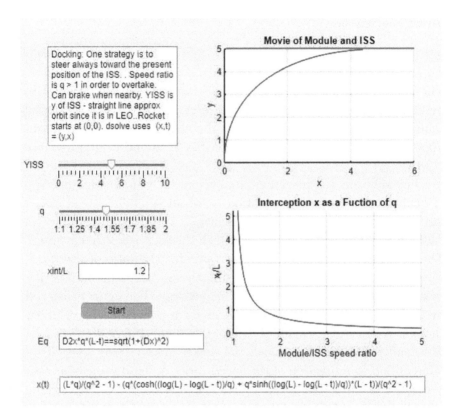

Figure 4.11: The rendezvous of a rocket with the ISS. The ISS altitude is L or YISS and the speed ratio of the rocket to the ISS is q. The symbolic solution for $x(t)$ is displayed in "dsolve" variables, which are $y(x)$ in this specific problem and plotted as a "movie". The first term in $x(t)$ is the x interception point.

that at $y = 0$, $v = 0$, a soft landing on the surface. The acceleration a is assumed to be constant, although, Eq. (4.12) it would increase as fuel is expended. However, it is assumed here, for simplicity, that $T - t \sim T$. The attraction of the Moon is assumed to be a constant acceleration, g, as the fall starts close to the surface. With those approximations, an analytic solution is possible if $a > g$.

$$t_r = t_{ff}[g/(a - g)], \quad t_{ff} = \sqrt{2y_o/[(a - g) + (a - g)^2/g]} \quad (4.23)$$

The App "Lunar_Lander" is used to show a movie of the lunar descent in Figure 4.12. The choice of a is made by the Slider. Looking at the

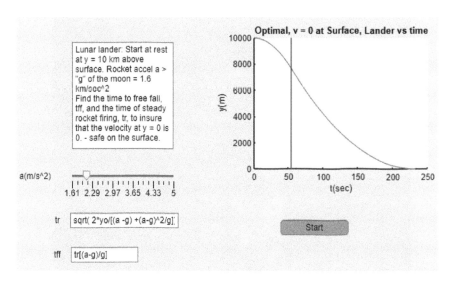

Figure 4.12:   Plot of the lunar descent choosing $y_o = 10$ km with Lunar Module acceleration, a, chosen by the Slider and starting at the location of the vertical red line when free-fall motion ends.

plot it is clear that the firing of the descent rocket has a smooth effect on the module.

A photo not of the descent but of the ascent as taken from the Apollo Command Module is shown in Figure 4.13. The Lunar Lander is ascending toward docking with the Earth in the background.

After this landing and exploration, the passengers will, presumably, need to return to the Earth since that is the unique place that humans have evolved. However, there are some issues because for a descent to the Earth's surface the atmosphere of the Earth must be traversed and the kinetic energy of the Command Module must be dispersed in order to make a soft landing but without sufficient fuel to make a powered descent. If there were a powered descent, a strategy like that for the Lunar Lander could be adopted.

4.11. Earth's Atmosphere

Before attempting to return to the Earth, a few digressions are made in order to define the atmosphere of the Earth and a few other bodies

Figure 4.13: Photo of a Lunar Lander launch in progress. The upper stage has been left on the Moon and the smaller module that is used to escape the gravitational field of the Moon and return to Earth is viewed from the Command Module which will be used to return to Earth, seen here as the "blue marble".

and explore aspects relevant to the landing. Some basic parameters for the atmosphere of the Earth are tabulated in Appendix E.

An App used to explore atmospheres is "Atmospheres". The effect of temperature, T, and molecular weight, A, of the atmospheric gases can be studied using the Sliders to set their values. The mean Maxwell–Boltzmann kinetic energies and the velocities of the molecules with this A and T are written using Edit Fields. The energy units are eV, which is appropriate at these temperatures (Figure 4.14).

The distribution of energies depends only on temperature, as in Eq. (2.16). The Maxwell–Boltzmann distribution for velocity is simply derived from Eq. (2.16), where the classical distribution of energies was defined. Changing the variables, the factor is $d\varepsilon/dv = mv$, or $dn/dv = dn/d\varepsilon(d\varepsilon/dv)$. The resulting velocity distribution

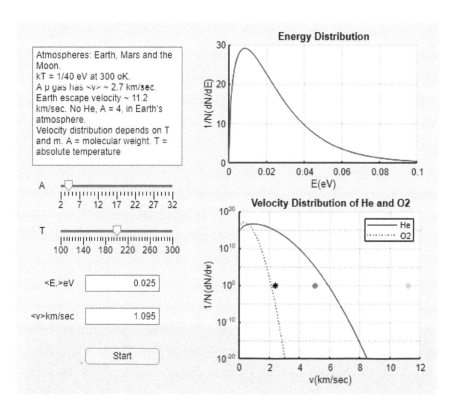

Figure 4.14:   Output of the App for specific values of A and T. The Edit Fields display the mean values of energy and velocity. The plots display, at T, the energy distribution and the velocity distributions for helium and O_2. The stars indicate escape velocities for Earth, Mars, and the Moon (descending order).

has factors $\sim v^2 \exp(-mv^2/2kT)$ and is generated and plotted for both helium and oxygen and compared to the escape velocities of Earth, Mars, and the Moon. It is clear that the Moon should be airless and that Mars will have a tenuous atmosphere at best. The mean energy is $3kT/2$ with a mean square velocity of $(3kT)/m$.

Consider the molecular composition of the Earth's atmosphere. Although helium is very abundant in stars, it is very rare on Earth. That is because its velocity distribution on Earth has thermally fluctuated over time into escape velocity. Indeed, there is a helium shortage on Earth just now due to the increased demand for it caused by the rapid expansion of superconducting applications. The existing sources of helium on Earth are from gas sequestered in geological structures.

The Earth has both an atmosphere and a magnetic field. The field traps charged incoming particles and transports them to the poles (*Aurora Borealis*). Together, they limit the radiation exposure for most of the surface.

On the surface, the exposure is mostly due to muons that only ionize materials since the atmosphere provides about 10 nuclear absorption lengths which absorb most of the nuclear cascade that occurs when the cosmic ray proton or nucleus interacts in the upper atmosphere. A schematic diagram of the nuclear cascades is shown in Figure 4.15. It dramatically illustrates the protection afforded by the atmosphere. Astronauts will not be adequately protected.

The surface annual exposure on Earth is about $3\,\text{mG/year}$. One Grey (G) is an equivalent energy deposit of $1\,\text{J/kg}$, or about 100 rad.

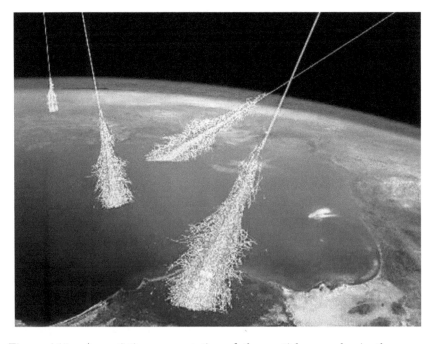

Figure 4.15: An artistic representation of the particle cascades in the upper atmosphere caused by incoming cosmic rays interacting there. It provides a graphic sense of the protection afforded by the Earth's atmosphere.

The dose for long-term health from additional sources is thought to be about 0.1 rad/year, although for 1 year, exposure of up to 5 rad is deemed to be acceptable for "radiation workers".

The absorbed energy is in Gray while the dose equivalent, which differs because different radiations produce different biological damage, is defined in Sievert (Sv). One difference arises because heavily ionizing radiation may be more damaging because the energy is released in a smaller physical region. For example, an estimate for cosmic rays in space is that the absorbed dose is only 0.2 Gy/year, while the dose equivalent is $\sim$1 Sv/year due to heavily ionizing iron nuclei in cosmic rays and the Z^2 scaling for ionization, where Z is the atomic number.

Humans have evolved in an environment provided with air, water, power from the Sun, and the protection of the atmosphere. On Mars, with a more tenuous atmosphere and no magnetic field, the exposure is estimated to be about 0.15 Sv/year, 15 rad/year, while during a 1-year trip to Mars and back, a dose of up to 1 Sv or 100 rad might be encountered. This radiation exposure is well in excess of that thought to be acceptable for radiation workers on Earth.

Recently, an astronaut spent a year LEO at the ISS. He had a twin who stayed on Earth, so that close comparisons could be made. Radiation-induced chromosome instability and other deleterious effects were observed compared to the twin who remained on Earth. This fact does not bode well for long-term trips beyond the Earth unless specific steps are taken to reduce the radiation exposure. Unfortunately, shielding from cosmic rays will add considerable weight to any spacecraft. This is a problem that needs to be solved in the near future if a Mars trip is planned.

4.12. Mach Cone

The atmosphere supports pressure waves, or sound waves, with a particular velocity. The speed of sound is approximately 0.32 km/s while the mean thermal speed is 0.39 km/s. Disturbances caused by the passage of a body through the atmosphere are quite different depending on whether the speed of the body exceeds the speed of sound or not. This behavior is familiar in the passage of boats and the water waves they create. The "bow wave" that a boat creates in its wake is a common experience for most people.

The same situation is seen in the case of a plane or rocket. The angle of the shock wave, or "Mach cone", is $\sin(\theta) = v_s/v$ for velocity v and sound velocity v_s. There is no shock wave for low velocities since formally the sine is greater than 1 then, but physically what is meant is that the individual disturbances do not add up. When the velocity of the emitter is equal to the velocity of sound, the wave is at 90° to the direction of travel. At very high emitter velocities, the emission angle shrinks to zero.

The parameters of the Mach cone are defined in the App "Mach_Cone". There are six emission times at equally spaced x positions from 0 to 5. The velocity of the sound waves is defined to be 1. A Slider allows the user to choose velocities both below and above the sound velocity. Mach cone angles from the threshold at 90° to about 20° can be generated. A specific result is shown in Figure 4.16. The green points on the x axis are the location of the

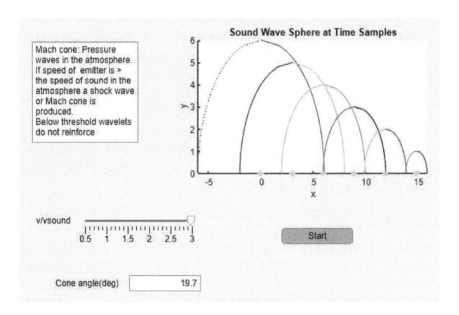

Figure 4.16:   Plot of the spheres of sound waves after emission at 6 equally spaced times (0, 5). The location of the emitter at those times are displayed as green dots.

emitter at the emission times. The lines are the locus of outgoing sound from the emission for both forward and backward emission. In this specific case, the Mach cone is very obvious. The user is encouraged to generate different combinations. There is effectively a short movie with only 6 frames for visual clarity.

A wind tunnel picture of the profile of the space shuttle is shown in Figure 4.17. The Mach cones that appear during supersonic re-entry are very visible. In this case, there is no pointlike emitter but the nosecone and the leading edges of the wings all contribute to the full shock wave pattern.

Figure 4.17: Wind tunnel visualization of the shock waves generated by the space shuttle during supersonic re-entry from LEO. Both the leading edge of the nose and the leading wing surfaces contribute strongly to the full shock wave.

4.13. Re-entry — Atmosphere

There is the question of how to return to Earth. It is assumed that the descent is unpowered and that the atmosphere will be used to brake the descent. One issue is that the air is compressed because it cannot get out of the way of the descent fast enough since the descent speed from LEO, about 8 km/s, is initially much greater than the speed of sound, which is 0.32 km/s at sea level, and standard temperature and pressure, STP. That means the air will heat up, much like the situation when a bicycle tire heats up when pumped to higher pressures.

This is a complex process, which can be approximated, for a first look, as a process of adiabatic compression. The heat flux, in W/m^2, is F_H and depends on the velocity of the descent object, v, and the density of the atmosphere, ρ, as

$$F_H = \rho v^3 / 3 \qquad (4.24)$$

As a numerical example, consider a 1,000-kg (1 tonne) descent module moving at LEO. The kinetic energy of the module is initially 6.4×10^{11} J or 640 GJ. The atmospheric falloff with altitude is assumed to be exponential with scale 9.14 km, so that the density at 100-km LEO is 2.1×10^{-5} kg/m^3. The heat flux is then 5.4 MW/m^2 which needs to be dissipated using a "heat shield" which basically is ablated and protects the module itself. A schematic picture of how that heat might be removed from the descent module is shown in Figure 4.18. For comparison 1 tonne of dynamite releases only 4.2 GJ of energy.

The kinematics of the descent are explored in the App "Orbit_Reenter". The thermodynamics are complex and will not be explored further. The order of the magnitude of heat flux is daunting enough. The rocket ascent was powered and was essentially a one-dimensional problem. The descent is unpowered and, by necessity, is a two-dimensional problem. The acceleration, Eq. (4.25), has a term due to the Earth's attraction and a drag term similar to what was seen during ascent, Eq. (4.19), but without the rocket power term (Figure 4.19):

$$d^2\vec{r}/dt^2 = -(GM_E)(\vec{r}/r^3) - [\rho(y)C_D(M)Av\vec{v}]/2 \qquad (4.25)$$

The drag term represents the atmosphere with a density, $\rho(y)$, which falls off exponentially with altitude and a term which depends on the Mach number, M, which is defined to be the ratio of the

Figure 4.18: Schematic view of the descent of Apollo Command Module astronauts when entering the tenuous upper Earth atmosphere. The orientation of the heat shield is critical for a successful re-entry.

speed of the object to the speed of sound. This behavior is simplified in the App, but the coefficient still peaks at an M value of 1. The drag deceleration depends on the area, A, of the object, points in the direction of the velocity, and is proportional to the square of the velocity.

The two-dimensional equations are solved numerically using the Matlab utility "ode45". The user can choose Slider values for the LEO altitude and the angle of descent. Since the thermal issues are being ignored, the angle is not very sensitive, but in reality, it is a parameter that needs to be finely tuned as several descent tragedies have illustrated. The results of a particular choice of parameters are displayed in Figure 4.19. The curves of altitude vs. velocity and altitude vs. deceleration are very characteristic of an unpowered descent.

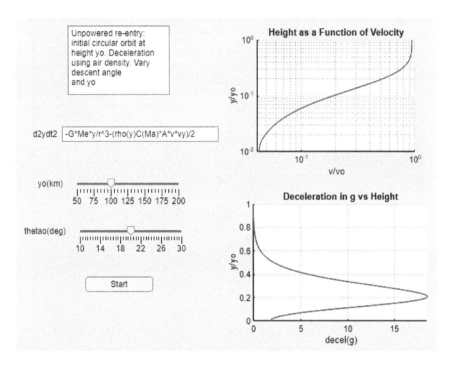

Figure 4.19: Results of the App "Orbit_Reenter" showing deceleration and the velocity of descent as a function of the altitude. The shapes of the curves are very distinctive of an unpowered descent.

The descent starts with a rapid loss of altitude but little loss of velocity since the atmosphere is attenuated at $100\,\text{km}$. At an altitude of about $50\,\text{km}$ and below, the loss of altitude with velocity is roughly linear. The final conditions $v = 0$ and $y = 0$ are not imposed, since the behavior is entirely determined by the initial conditions in this model. In practice, small rockets exist to make adjustments so that it is not simply an initial value problem but a problem in piloting.

The deceleration is small initially and then peaks at an altitude of about $20\,\text{km}$ to a value of almost 20 "gees". After the peak, the falling value of the velocity means that the deceleration falls rapidly to zero at lower altitudes. Typically, the landing is enabled using parachutes to come to rest on either land or a body of water. These techniques give the pilot a margin for error.

4.14. The Space Elevator

So far in this chapter the use of single-stage and multi-stage rockets
has been shown to be practical for LEO such as is required to
construct the ISS. Larger rockets have taken people to the Moon to
land there and return to the Earth. These voyages were accomplished
about 50 years ago. Is there an easier way? Science fiction writers
have posited several. In Sections 4.15 and 4.16 two such ideas
applicable to travel near the Earth are explored from the point of
view of the physics and engineering issues.

Many satellites are placed in the geosynchronous Earth orbit
(GEO). Communications are simplified if the satellite is always
above the same point on the Earth. The balance of attraction and
centrifugal repulsion for an angular velocity ω was shown in Eq. (4.4):

$$a = -GM/r^2 + \omega^2 r$$
$$r_{\text{GEO}} = (GM/\omega^2)^{1/3}$$

$$(4.26)$$

This balance makes it possible to place satellites in orbit, but
there could also be structures standing on the Earth. Using the
angular velocity of the Earth, the equatorial radius for GEO is about
42,000 km or about 35,800 km in altitude. The lowest orbit occurs at
the equator.

The idea is then very "simple". One builds a tower and at leisure
climbs it or rides an elevator to a platform at that altitude. At the top
there is no acceleration and one is "weightless". Since the platform
is fixed with respect to the ground, no messy rocketry is needed.
An ISS is assembled just by climbing enough stairs at the desired
pace. What could possibly go wrong? Let us look at the necessary
materials to use in the construction of this "space elevator". Even
the tower of Babel was wider at the base than the top.

The area, A, of the elevator tower should decrease as the height
increases because the lower portions need to support the weight above
them under compression. In this simplest realization, without an
additional counterweight extending further in radius, the cable is
under compression. The cable must be under a pressure less than
the tensile strength of the material. The tensile strength has units

of pressure and is denoted by S while the density is ρ. The unit of $\rho GM/r$ is also the same as that of pressure. One possible design criterion is to keep all elements of the elevator cable under constant stress. The dependence of the cable area on r in this case is solved symbolically, with the approximate result for r_E much greater than r at the elevator top

$$dA/dr = A[\rho GM/(Sr^2) - (\rho\omega^2 r)/S]$$
$$A_{\text{GEO}}/A_{\text{earth}} \sim \exp[-(\rho/S)(GM_E/r_E)]$$

$$(4.27)$$

At the GEO altitude, there is no "weight" because the forces balance. The numerical estimates are made in the Matlab script "space_elevator". Printout made by the script is shown in Figure 4.20. The minimum area ratio exponent for titanium, with the best strength to weight ratio of the metals, is 591 which is enormous. A material with low density, such as Kevlar, has a lower exponent, "only" 25.

```
>> space_elevator

AA =

exp(C3 - (rho*(r^3*w^2 + 2*G*M))/(2*S*r))

Height of Elevator = 35863.3 (km)
Ratio of cable area from Earth to Elevator - Approx
  exp((rho/S)*(GM_E/r_E))
Area Ratio Exponent for Titanium = 591.228
Area Ratio Exponent for Kevlar = 25.1873
```

Figure 4.20: Printout of the Matlab script for a space elevator.

A plot for possible man-made low-density materials with density defined to be that of water is shown in Figure 4.21. Reasonable area ratios are not obtained until the strength exceeds tens of GPA. That stress level constitutes a severe engineering issue.

A beautiful idea founders on a simple engineering issue with the strength of materials. A supply of "unobtainium" is needed to build the elevator. Perhaps the progress in man-made materials will, in the

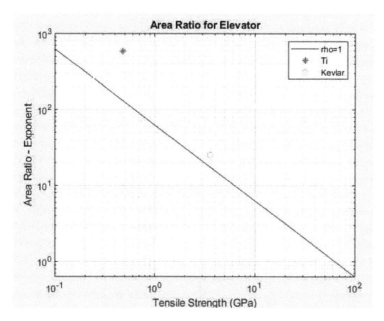

Figure 4.21: Plot of the area ratio exponent for density 1, in gm/cm^3, or 10^3 in kg/m^3, materials as a function of tensile strength. An area ratio of about 400 requires a tensile strength of about 10 GPa.

future, enable the manufacture of such materials. Also, in a less deep gravity well, the problem would be exponentially easier so perhaps elevators on low-mass asteroids or moons might be feasible.

A schematic diagram of a possible variant elevator is shown in Figure 4.22. The idea is to have a counterweight so that the cable is under tension rather than compression. That is an added complication, but has the advantage that the maximum stress appears at the location of the GEO.

4.15. The Rail Gun

Another speculative device for achieving orbits in the solar system is the rail gun. Such devices have been developed on a small scale as weapons. In science fiction stories, they are posited to be used on the Moon or other atmosphere-free asteroids and the like. Unlike rockets, there is no fuel to carry. There is a structure with a sliding

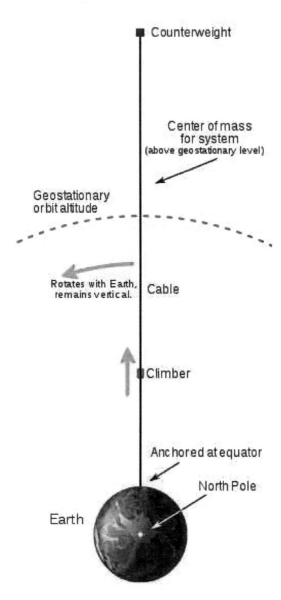

Figure 4.22: Schematic diagram of a possibly different "space elevator". The cable profile is clearly not drawn to scale.

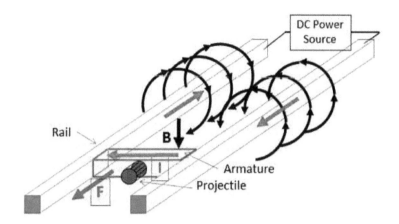

Figure 4.23: Schematic diagram of a rail gun. There are two parallel rails carrying a current which runs through the armature. The Lorentz force, F, is in the direction of the rails and propels the projectile.

armature on rails carrying a current I. A schematic diagram of such a structure is shown in Figure 4.23. The current is shown as green arrows and the magnetic fields they create as black circles.

Imagine such a structure on the Moon. It would be used to launch materials into orbit or toward the Earth. The "projectile" in Figure 4.23 is the payload. It would be released at the end of the rails. The B fields and the force on the armature are estimated using Ampere's law. The force on the armature follows from the Lorentz force applied to the current I and is approximately given as

$$B \sim [\mu_o I/(2\pi)][2/b],$$
$$d\vec{F} = Id\vec{L}x\vec{B}, \quad F \sim \mu_o I^2/(2\pi)$$

(4.28)

The current is I while the rail separation is b. The armature length is $L \sim b$. The impulsive force is integrated for a total mass of payload plus armature of m. The current is driven by a charged capacitor of capacitance C, charge Q with stored energy $U_c = Q^2/(2C)$. Assuming a small inductive time constant, the current rises instantaneously to an initial maximum value of $I_{\max} = Q/(RC)$, where R is the resistance of the conductive armature and the current decays with the time constant $\tau = RC$. The rails are assumed to be superconducting, with no resistance. Integrating the squared current over time, the final velocity of the armature plus payload is given as

$$v_f \sim (\mu_o/\pi)(Q^2/RC)/(m_a + m_p) \tag{4.29}$$

There are several geometric and practical coefficients of order 1 or less which are not specified in this very oversimplified treatment. The App "Railgun_Moon" defines the parameters needed to launch a payload to escape the Moon's gravity. A sample output is given in Figure 4.24. The Sliders are used to choose the energy stored in the capacitor bank and the length of the armature. The RC time constant and the final velocity are displayed using Edit Fields, while the time development of the armature current and the velocity are plotted. It is assumed that the armature does not move appreciably while the current pulse is active: the force is impulsive. In this specific case, the current pulse lasts less than 1 μsec discharging about 6 MJ into a 1-m-long copper armature.

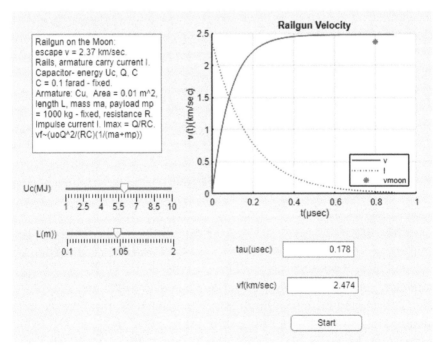

Figure 4.24:   Output figure for the App "Railgun_App" for a choice of parameters such that escape velocity for the Moon is just achieved.

Of course this is not a method for personnel given the enormous g forces. There are many, many, technical issues. The rails repel one another. There is heat generated by the pulses, etc. Nevertheless, the rail gun is not outside the realm of possibility especially for shipping freight from an airless or low-gravity body, such as the Moon.

4.16. Tides and Moons

The tidal forces are the intrinsic aspects of gravity. The tidal forces in Newtonian gravity have already been seen in the discussion of tidal stresses near a black hole. Specifically, the Moon exerts a tidal force on the Earth which causes tides to rise in the liquid seas. They occur twice a day. The force as seen before in Eq. (3.20) causes the tides at the surface of the Earth, radius R_e, is F_T where R_{em} is the Earth–Moon distance. It is derivable from a potential, Φ_T which scales as $F_T R_e$ at the surface. In this case, M is the mass of the Moon. Assuming the water surface is at a gravitational equipotential, the tidal height is then h, calculable in terms of basic Earth–Moon constants. Numerically, h is 0.56 m which is close to the observed tides, although the actual height depends on the local environment of the water:

$$F_T/m \sim 2GMR_e/R_{em}^3$$
$$gh \sim \Phi_T \sim (F_T R_e)/m \qquad (4.30)$$
$$h = 3GMR_e^2/(2gR_{em}^3)$$

Farther away from home, space travelers may wish to visit the moons of the planets. What limits the size or radius of planetary moons? There is a "Roche limit" on how close a moonlet can approach a gravitating body like the Earth. Tidal forces will pull a close satellite apart. The limit on the orbit radius, R, of the moonlet with a mass m and radius r is, for a central attracting mass M, assuming the moonlet is only held together by gravity occurs when the tidal force exceeds the self-gravity of the moonlet. This is called the Roche limit:

$$(2M/m)(r/R)^3 = 1 \qquad (4.31)$$

The limit depends only on mass and distance ratios because the tidal force is proportional to G but the force holding the moonlet together is also. A schematic diagram of the tidal disruption of a moonlet is shown in Figure 4.25. There are no known moons in the solar system that violate this limit so that solar system explorations will not be able to explore moonlets of solar planets. However, astronomers have observed tidal disruptions of entire stars passing near a supermassive hole at the center of the galaxy and reported in 2016. Closer to home, the rings of Saturn are inside the Roche limit and thus may be the remnant of a disrupted moon.

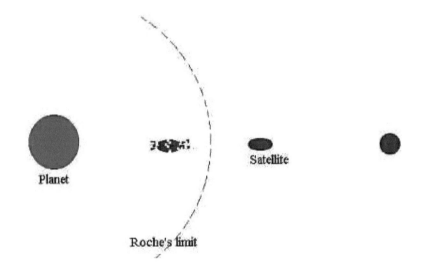

Figure 4.25:   Schematic diagram of the disruption of a moonlet by the tidal forces of a more massive planet.

The tides and the Roche limits are displayed using the App "Tides_Roche" with output shown in Figure 4.26. A plot of the tides, greatly exaggerated, is created by the App. Sliders are used to choose the ratios of radii and mass for the planet and the moonlet. The Roche limit is shown numerically with an Edit Field. A plot shows the limit boundary for R/r and M/m and three specific points showing solar system moons. Mercury and Venus have no moons, so the innermost moons of the Earth, Mars, and Jupiter are plotted.

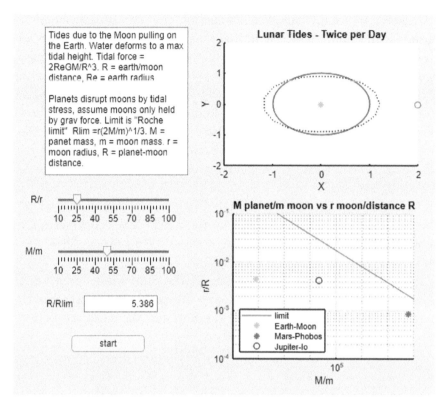

Figure 4.26:   The effect of tidal forces on the Earth on bodies of water and on possible tidal disruption of moonlets in the solar system obtained using the App "Tides_Roche". No known moon exceeds the limit.

4.17. Lagrange Points

The discussion of orbits started with the ideal case described in Eq. (4.6) with a very massive Sun, which remained fixed at the origin of coordinates. This simple approximation was changed when binary systems were described using Eq. (3.14) and the CM frame was defined. In general, a two-body system can be reduced to a one-body system for orbits even when the masses are not very dissimilar. The system acts like a single particle with a reduced mass, $\mu = M_1 M_2/(M_1 + M_2)$. Gravitationally, the system acts as a mass μ and a momentum and angular momentum of magnitude appropriate to a mass μ. In Sections 4.18 and 4.19 the Kepler approximations will be loosened in order to explore some of the consequences.

There are many space missions for the use of astronomical observatories that are placed at "Lagrange points". They are a special case of the "three-body problem" in this case where a small mass orbits a large mass and a test body also orbits but without relative acceleration. All three bodies rotate in lockstep. For a system of the Earth and the Moon and a small body, we will look at the three points on the Earth–Moon axis where the relative position of the body and the Moon is fixed. These points provide a stable platform for space observatories. A second set occurs for the Sun–Earth — test body system.

The stable points can be evaluated in a co-rotating system where the two masses are at fixed points on the x axis. The coordinates are shown in Figure 4.27. In this case, $a = -x_1 + x_2$. $M = M_1 + M_2$, and $\omega^2 = \mathrm{GM}/a^3$.

The Earth is at a distance R from the Sun. First, for a body between the two and at a distance r from the Earth, the Earth partially cancels the attraction of the Sun and therefore the period

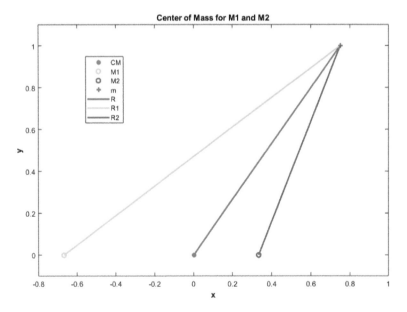

Figure 4.27: Schematic diagram of the CM at the origin for 2 masses located at $(-x_1, 0)$ and $(x_2, 0)$ in a co-rotating frame rotating with angular velocity ω, where the masses are at rest.

of the body increases and matches that of the Earth so that the body is fixed in space with respect to the Earth. In general, the velocity goes as the inverse square root of the distance from the mass while the period goes as the 3/2 power, as already seen in Eq. (4.5):

$$v = \sqrt{GM/r}, \quad v/c = \sqrt{R_s/2r},$$
$$\tau = 2\pi r^{3/2}/\sqrt{GM} = 2^{3/2}\pi(r/c)\sqrt{r/R_s} \tag{4.32}$$

Assuming the period of the Earth is fully defined by the mass of the Sun, the stationary L_1 position is given as

$$1/R^3 = 1/(R-r)^3 - q/[r^2(R-r)], \quad q = M_E/M_S \tag{4.33}$$

This result would apply to the Earth–Moon system also. The second Lagrange point, L_2, occurs with the body beyond the orbit of the Earth. In that case, the Sun and Earth reinforce each other and the period decreases to match that of the Earth. The implicit solution for r is then

$$1/R^3 = 1/(R+r)^3 + q/[r^2(R+r)] \tag{4.34}$$

The third Lagrange point, L_3, is also along the line of the Sun and Earth but has the test body opposite the Earth with the Sun in between:

$$1/R^3 = 1/r^3 + q/[r(R+r)^2] \tag{4.35}$$

These implicit formulae do not have a simple closed-form solution for r. However, taking q and r/R to be small, the L_1 and L_2 points have approximate solutions $r/R \sim (q/3)^{1/3}$ with L_1 nearer the Sun than the Earth and L_2 on the far side at the same distance. For L_3, r is $\sim R$.

A more visual way to see the Lagrange points is to use a potential and look for the potential minima. The potential due to the Sun, M_1, at $(-x_1, 0)$, the Earth, M_2, at $(x_2, 0)$, the CM at $(0,0)$, and the centrifugal potential of the orbit about the CM is shown in Eq. (4.36), for the general case. The centrifugal potential scales as radius squared, as seen previously in Eq. (4.4). The masses M_1 and

M_2 are at distances R_1 and R_2 from the observation point (x, y) located at the test mass:

$$x_1 = M_2/(M_1 + M_2), \quad x_2 = M_1/(M_1 + M_2)$$
$$a = x_1 + x_2, \quad \omega^2 = G(M_1 + M_2)/a^3$$
$$\Phi(x, y) = -G[M_1/R_1 + M_2/R_2] - (\omega R)^2/2 \tag{4.36}$$
$$R_1^2 = (x + x_1)^2 + y^2, \quad R_2^2 = (x - x_2)^2 + y^2$$

The Lagrange potential for a two-body system is explored using the App "Lagrange_Earth_Moon_2". It applies to any system with two major masses, such as the Earth–Moon. The figure created by the App is shown in Figure 4.28. The Slider is used to choose the

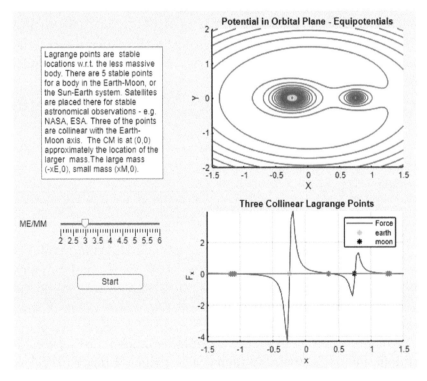

Figure 4.28: Plots of the potential and the force along the x axis which is the line of center of the two masses. The locations of L_1, L_2, and L_3 are visible as the locus of red stars at $Fx = 0$. The locations of L_4 and L_5, which are at non-zero values of y, are not as obvious and have not been well established here.

mass ratio of the two bodies. It is initially set to small values to make the equipotentials more visual.

The user can change the location of the Lagrange points by changing the value of M_E/M_M. The non-collinear Lagrange points are not well displayed but a minimum does exist off the x axis. The Matlab utility "gradient" applied to the potential $\Phi(x, y)$ is used to find the forces F_x and F_y. The $F_x = 0$ points are found using the utility "sort" on $|F_x|$. Multiple points are plotted since the "sort" is not very accurate on the F_x vector when it is near zero at the L_2 and L_3 locations.

For the Earth–Moon system, with $M_E/M_M = 82.2$, the L_1 and L_2 points are about 15% more or less than the Earth–Moon distance. For the Sun–Earth system, the distances are only about 1% different owing to the larger mass ratio. Many observatories such as Planck and WMAP are located at L_2 for the Sun–Earth system. The Lagrange points are also natural places to find asteroids, for example, the "Trojan" L_4, L_5 points of the Sun–Jupiter system.

4.18. Planetary Perturbations

The approach to individual planetary orbits so far has been idealized. The Sun was taken to be fixed in space while the planets were taken to be test masses in independent elliptical orbits about the Sun. That model is only true for zero-mass planets. The CM of two masses, one at a distance R_M from the CM and a second mass m at a distance R_m acts like a system of reduced mass μ in a frame where the CM is at rest at the origin. If M is much greater than m, $R_M \sim 0$, $R_m \sim R$ and $\mu \sim m$, thus recovering the previous approximations. The other extreme case is for equal masses as was discussed in the treatment of binary systems. In that case, $R_M = R_m = R/2$ and $\mu = M/2$:

$$R_M = [m/(m + M)]R, \quad R_m = [M/(m + M)]R$$
$$R = R_m + R_M, \quad \mu = Mm/(m + M)$$

$$(4.37)$$

The Lagrange points are a simple limit of the three-body problem because the spacecraft does not affect the Earth or the Sun, which orbit about their common CM. Another simplified three-body problem occurs when one mass, the Sun in this case, is much more massive than any planet. In that case, the perturbations are between the planets. For the solar system, Jupiter has an orbit at 5.2 AU and is 318 times more massive than the Earth. However, it is still about one thousand times less massive than the Sun, which justifies the approximation that the Sun is at the CM and at rest.

This special case is explored in the App "Jup_Earth_Sun_App". The mass and radius of "Jupiter" are chosen by the Slider. The period of "Jupiter" is displayed in an EditField. The equations of motion for the forces between Earth and the Sun and "Jupiter" and the Sun supply the unperturbed orbits using "kepler3" as an "ode45" special case. The perturbed orbits arise when the force between the Earth and "Jupiter" is also turned on and the results of a numerical integration are computed and displayed.

A movie of the orbits is made for about five orbits of the Earth. Finally, the deviation of the Earth orbit from the unperturbed case is shown graphically. Circular unperturbed orbits are assumed. The period of Jupiter is 11.8 years as displayed in an EditField at startup. The effects of Jupiter alone are a few percent per year. An extreme case is shown in Figure 4.29, where the mass of "Jupiter" is increased by a factor of 50 and the orbit is reduced to a radius of 2 AU with a period of 2.9 years. The orbit of "Jupiter" in this case is still changed only a little, while that of the Earth is pulled off its circular path by a noticeable amount with a maximum change of about 20%. The user is, as always, encouraged to adjust the Sliders and observe the results of the change.

For a realistic travel between the planets, the small perturbations will necessarily need to be taken into account. The simple Keplerian model that all planets are massless will not suffice for true navigation in the solar system. In Sections 4.19 and 4.20 travel to Mars and Jupiter, and thence escaping the solar system entirely, will be explored. For Jupiter, in particular, the gravitational field will prove to be crucially useful.

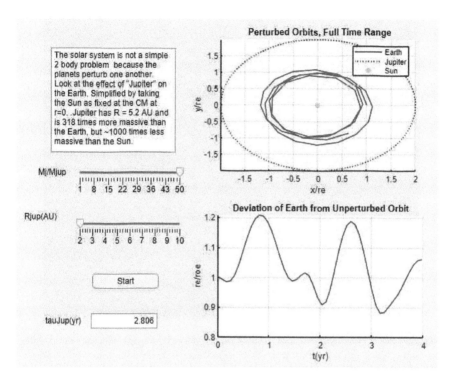

Figure 4.29: Output of the App "Jup_Earth_Sun_App" showing the perturbed orbit of the Earth for a "Jupiter" whose mass and orbital radius are chosen by the Slider.

4.19. Mars–Earth — Transfer Orbit

Having explored LEO and the Moon, the next step would be a trip to a nearby planet, such as Mars. What rocketry does that entail? The basic idea is to exit the Earth orbit, assumed to be circular, enter an elliptical orbit with a velocity change, and then enter the orbit of Mars with a second burn.

This "Hohmann" orbit is an elliptical orbit to transfer between two circular orbits. This plan is popular because it uses the minimum amount of energy. A schematic diagram of such a transfer is shown in Figure 4.30. The launch is at the major axis of Earth and the arrival is at the location of Mars 180° away from Earth at launch.

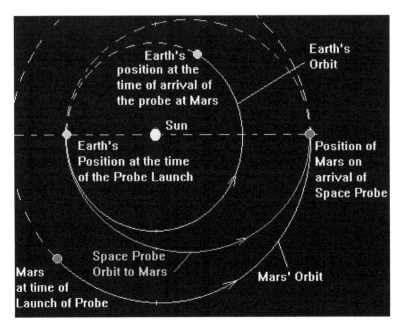

Figure 4.30:　Positions of Earth and Mars during a Hohmann transfer orbit. The necessity for a launch window is clear since Earth at launch must be opposite Mars at arrival. The trip takes about 9 months and the launch window is only available every 26 months.

This launch window is not always available and the time must be chosen.

The analysis of the transfer orbits follows from the general Kepler analysis, Eq. (4.5). The period of an orbit depends only on the 3/2 power of the semi-major axis, a. The energy, U, of the orbit is a constant of the motion and depends only on the inverse of a:

$$U/m = v^2/2 - GM/r = -GM/(2a)$$
$$v^2 = GM(2/r - 1/a), \quad (v/c)^2 = R_s(1/r - 1/2a) \tag{4.38}$$

From these results, one can find the velocity difference needed to go from a circular orbit with radius R_1 to an elliptical orbit for semi-major axis $R_1 + R_2$. The velocity difference to then inject into a circular orbit at R_2 and the time it takes, the half period of the transfer ellipse, τ_H are given as follows:

$$\Delta v_1 = \sqrt{GM/R_1}\left(\sqrt{2R_2/(R_1 + R_2)} - 1\right)$$
$$\Delta v_2 = \sqrt{GM/R_2}\left(-\sqrt{2R_1/(R_1 + R_2)} + 1\right) \qquad (4.39)$$
$$\tau_H = \pi\sqrt{(R_1 + R_2)^3/GM}$$

The trip requires two "burns", one to go from the velocity at the Earth's radius, R_1, to the Hohmann ellipse and one to go from the ellipse a half period later into the target orbit, R_2. A one-way trip varies from 0.71 years to Mars to 16 years for Uranus. Given the radiation issues and the sheer time scale, these trips appear to be limited, at least for human passengers, to Mars at most.

These trips are explored in the App "Outer_Planets". The first plot generated is a "movie" of the orbits of the four inner planets, Mercury, Venus, Earth, and Mars. The movie is for 1 Earth year, and Mars only completes about a half a period.

The other plot shows orbits from Earth to the outer planets. A particular output for the trip to Mars is shown in Figure 4.31. The planet is chosen using an EditField. The period, orbital velocity, and radius for the chosen planet appear in three EditFields. The transfer orbit semi-major axis, the trip time, and the two velocity changes are also shown in EditFields. For a Mars trip, the change in velocities at the beginning and end of the trip are only about 10% of the basic orbital velocity. Other, faster, transfer orbits are possible. They would minimize the radiation exposure to cosmic rays. However, the tradeoff is in the reduction of the mission payload fraction.

For reference to Section 4.20, a Hohman trip to Jupiter takes 2.7 years and ends at about 5 AU. The orbital velocity of Jupiter is 13.1 km/s.

4.20. Jupiter Gravity Assist

How does the exploration of the outer solar planets go forward? The Mars trip is already a stretch. The outer planet trips are of much longer duration. As noted in Section 4.19 on rocketry, the escape velocity for the Sun is about 42 km/s and the existing rockets barely

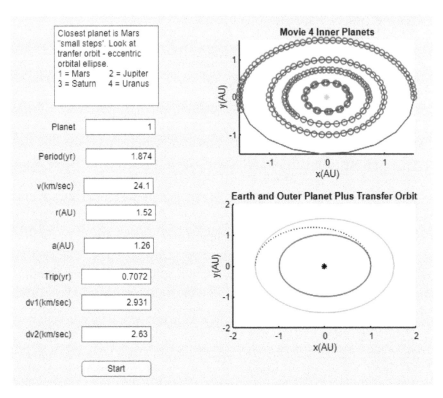

Figure 4.31: The App "Outer_Planets" is used to create the figure. The movie shows the 1-year orbits of the four inner planets. The specific Hohmann ellipse for a trip from Earth to Mars is shown in the second plot.

get a large payload into LEO. How can the solar system be explored near the outer planets?

In fact all the outer planets now have been explored by unmanned missions and, indeed, some missions have exited the entire solar system. The trick has been to borrow some momentum from the large outer planets such as Jupiter and Saturn. Orbits of the Voyager spacecraft are shown in Figure 4.32. They both used "gravity assists" from multiple large outer solar planets. They were thus also able to explore the outer planets Jupiter, Saturn, Uranus, and Neptune during their grand tour before leaving the solar system. Note that the missions still took 12 years to get to Neptune, even with a gravity assist. The Voyager missions have Hohmann-like trajectories,

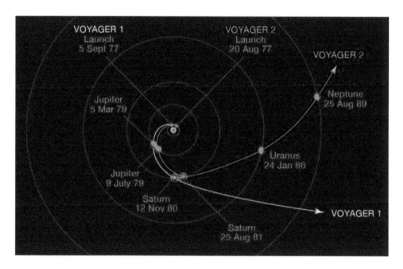

Figure 4.32: Orbits of the Voyager spacecraft. Both have since left the solar system entirely.

launched nearly opposite to the point of arrival at the orbit of Jupiter. The trip time to Jupiter was about 1.6 years, less than a trip time on a Hohmann orbit.

How was this possible? We already said the orbits have a conserved total energy, so how did the spacecraft gain enough energy to escape the solar system? It has already been hinted that the planets are not just massless test bodies, so that they can interact with other masses and not just orbit the Sun. In the simplified case of a one-dimensional collision of a light object with a very heavy moving object, energy and momentum are conserved even when the light object reverses direction and exits with a speed approximately twice that of the heavy object. In the gravity assist case, there is no collision, but there is an important interaction with the moving gravitational field.

The orbital velocity of Jupiter is about 13.1 km/s, and at a radius of 5.2 AU, the local escape velocity is only 18.5 km/s. A schematic plot of the velocity for Voyager 2 during the grand tour is shown in Figure 4.33. During the encounter with Jupiter, the spacecraft attained solar escape velocity, picking up about 12 km/s which is comparable to the orbital velocity of Jupiter. Subsequent gravity

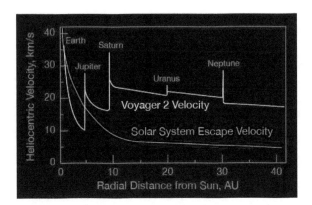

Figure 4.33:   Plot of the escape velocity as a function of distance from the Sun and the velocity of Voyager 2 during its grand tour.

assists allowed Voyager 2 to attain a velocity of about 18 km/s at a distance of about 40 AU, well beyond the local solar escape velocity at 40 AU of about 5 km/s.

To explore a bit deeper, the App "Jupiter_Gravity_Assist_2" can be invoked. The results for a specific choice of parameters are shown in Figure 4.34. The Sliders allow the user to pick the spacecraft velocity and the initial vertical angle of that velocity at some distance from "Jupiter". The Slider range is chosen so that the encounter with "Jupiter" can be chosen to either lead or follow the orbit of Jupiter. There are also settings where the spacecraft slows down. Indeed, a planet can also be used to slow down as well as speed up a spacecraft instead of using rocket fuel. Any planet that a spacecraft encounters can be thought of as a reservoir of momentum to be tapped.

The user chooses the spacecraft velocity. Subsequently, a "movie" of the spacecraft and "Jupiter" is made and the total trajectory is plotted along with a second plot of the evolution of the velocity components. The points are "frames" of the movie at equal times. The orbit of "Jupiter" does not change by construction while the spacecraft may slow or speed up. In this specific plot the increase in spacecraft speed is quite evident. The Slider values were chosen to give a final velocity with respect to Jupiter of about 1.4. The approximate terminal velocity with respect to the velocity of "Jupiter" is shown as an EditField.

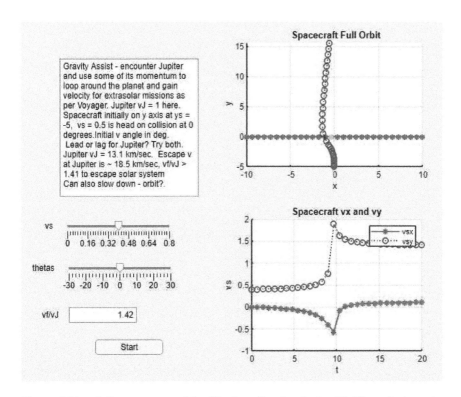

Figure 4.34: A figure generated by "Jupiter_Gravity_Assist_2". The velocity gain is about what would be needed to attain local escape velocity in a true Jupiter gravity assist maneuver.

A second plot shows the spacecraft x and y velocities as a function of time during the encounter. The user is encouraged to play with the initial spacecraft velocity so as to both lead and follow Jupiter and watch the changes in trajectory. The green dot shows the position of "Jupiter" when the spacecraft crosses $y = 0$. Typically arriving after "Jupiter", the spacecraft speeds up, while arriving before, it slows down. The gravity of field of "Jupiter" can slow down the spacecraft and even capture it in an orbit. These types of maneuvers can save on fuel.

Voyager 2 has escaped the solar system with a terminal velocity of about $17\,\mathrm{km/s}$. It will travel 5 light years (ly) about the distance to the closest stars, in about 87,000 years.

Chapter 5

Apps for Star Treks

"Across the sea of space, the stars are other Suns."

— **Carl Sagan**

"There must be someway outta here."

— **Bob Dylan**

5.1. Planets and Stellar Wobble

So far, only LEO has been explored by humans with the establishment of the ISS and many satellites. The Moon was briefly explored 50 years ago and then abandoned. All the other planets have had missions made by unmanned devices. Satellites at GEO are ubiquitous, and many scientific observatories exist at Lagrange points. Nevertheless, a manned Mars mission has not been accomplished and even then the radiation dose during the trip and the stay on Mars makes human travel problematic. Longer missions to the outer planets would only make the current technology more difficult to use. For example, communication from Earth to Jupiter is a two-way trip of about 8 AU or about 1 light hour. Conversations will be difficult.

Are there extra-solar planets to explore? One way to search for them is to look at nearby stars for orbital wobble or light fluctuations. The CM is stable but the star moves slightly in space due to the perturbations of the planets. The wobble size and period is a function of the mass of the star and the orbital radius and mass of the planet (CM motion, Eq. (4.37)). The CM is at rest at the origin in Figure 5.1 where the output figure of the App "Exoplanet_App" is shown for a particular planetary mass. The plot is first a "movie" of about

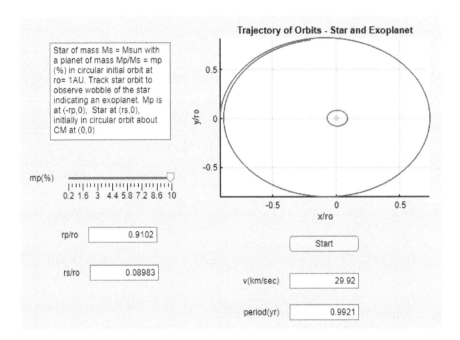

Figure 5.1: Examination of the effect of a planet on a star of solar mass at a of radius 1 AU. The mass of the planet can be estimated by using the Slider.

one planetary period. Then the entire trajectory of both the planet and the star are shown. The dynamics is solved for numerically using "ode45". The period of the wobble is essentially the planetary period. The first exoplanet was discovered in 1995. That discovery was recognized with the award of a Nobel Prize in 2019.

In addition to the stellar trajectory, the periodic velocity shift in the star can be observed by looking at the Doppler shift in the spectral lines of the star as it approaches and retreats from the observer. For example, the Sun has a velocity change of about 13 m/s due to Jupiter, a spectral shift which has been observed even though it is only a fractional shift of about 4×10^{-8}. If the star system is a binary one, there may be eclipses observed which then yield additional information of the masses and orbits of the full system.

Besides the motion of the stars, a direct observation of the passage of the planet between the star and the observer is possible. A schematic view of such an occurrence is shown in Figure 5.2. The transit time of the loss of star light and the details of how the light

Figure 5.2: Schematic representation of the transit of a star by an exoplanet.

loss falls off gives much information on the size of the intervening planet. A "transit" by a planet can be characterized by time duration, depth of light loss, and the shape of the edges (sensitive to planetary size). Absorption by the atmosphere of the planet may also indicate the type of atmosphere that the exoplanet possesses. In fact, a search for organic atmospheric molecules could even indicate the existence of life on the planet. A wealth of information has thus recently been obtained. In fact, there are many exoplanets known, some rather like Earth.

From the previous discussion of stars, for any given star the mass and radius can be estimated with reasonable accuracy using stellar models. The time a planet takes to occlude a star gives the planetary radius if the mass of the star is assumed so that the orbital velocity is calculable. In all, the radius and mass of the planet can be measured to some accuracy. There are additional tools such as gravitational microlensing which are also used. The great news is that extra-solar planets exist and they are, indeed, very common. A plot of the information on exoplanets is shown in Figure 5.3. Note that there is a bias toward large planets in close orbits because they

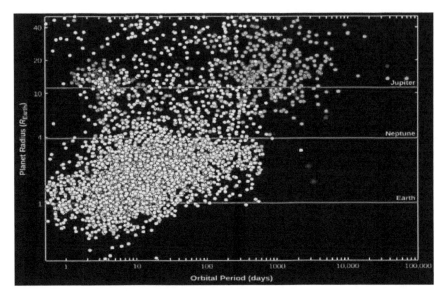

Figure 5.3:   Plot of orbital period vs. radius of the presently observed exoplanets where different methods of observation were used to obtain the summary results.

give larger signals for the search strategies that are used. As the methods of detection improve, that bias will be reduced. For now an Earth-sized planet with a year-long orbital period has not been observed. However, a rapid increase in the number of exoplanets is occuring. One exciting possibility is the absorption spectra of star light by the exoplanet and the search for spectra which indicate a possible Earthlike atmosphere. The new data assure us that there are exoplanets of many types which would be of enormous interest to explore, either by robotic missions or by human exploration.

5.2. Brute Force Approach

The exoplanets are out there. Is there a way to directly visit them and explore? Can this be done by passengers or only by instruments? The ratio of 1 light year (ly) to one AU is about 60,000. This implies that stellar exploration will be rather more difficult than solar exploration. Using existing technology trips of order of one hundred thousand years might be expected.

First, one can look at using rocketry since that technique is the standard solar method. One possibility is to assemble a large "ark" in LEO. In that case, the ark can be assembled without worrying about structural integrity on Earth, rather like the ISS. However, this model assumes numerous shuttle trips from Earth in order to ferry up all the needed materials. One way or another, it is costly to escape the gravity well of the Earth. It is possible materials may come from the Moon or elsewhere, perhaps the asteroids, but those sources would have their own costs.

For an ark assembled in LEO, the orbital velocity of about 7.7 km/s for a 300-km altitude can be used at launch time. A Hohmann trajectory to Jupiter needs an initial velocity change of 8.8 km/s and takes 2.8 years as can be seen using the App "Outer_Planets". Jupiter has an orbital velocity of 13.1 km/s for use in a gravity assist. Assume that the final ark velocity is 20 km/s at Uranus at 30 AU. The terminal trip velocity is reduced because the ark still needs to escape the attraction of the Sun moving outward away from Uranus.

Under these assumptions, the App "Ark_Ship" explores aspects of a trip of 5 ly. The solution for the rocketry during the launch phase from LEO is solved symbolically, and the acceleration, velocity, and position are shown as part of Figure 5.4. The acceleration is as before; the difference is only that there is an initial velocity, veo, ascribed to the LEO. Using Sliders, the user can select the exhaust velocity and the payload percentage. The trip velocity is defined to be the velocity after the ark has escaped the solar system. The trip time is approximated using the distance and the trip velocity and is displayed using Edit Fields.

The trip assumes a burn from LEO velocity of 7.7 km/s to Hohmann first-burn velocity of 8.8 km/s. The small increment is optimistic, but it implies a large payload fraction. This is indeed so, but it ignores the payload ratios needed to build the ark from materials found in the gravity well of the Earth. The trip takes about 81,000 years. Due to the duration an "ark" is assumed, one which imagines multiple, perhaps a thousand, generations living in the vehicle.

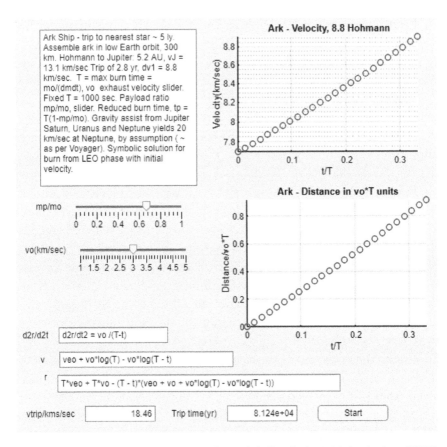

Figure 5.4:   Figure generated by the App "Ark_Trip". An initial velocity of LEO is assumed to start the trip. In this case, the payload is chosen so that the needed Hohmann velocity change is just attained with the chosen exhaust velocity.

Clearly, this interstellar rocketry is going to be a slow business. The velocities of travel all have a scale set by the exhaust velocity and that scale is approximately 4 km/s in the best case. One way to tackle the problem is to assume that the crew "hibernates" in some fashion. Cryogenic sleep is a popular science fiction trope. Even in that case, the issue of radiation exposure due to cosmic rays and the consequent genomic degradation remains. For example, long missions like Voyager have custom radiation "hard" electronics deployed precisely because of the radiation damage caused by cosmic rays. Humans are somewhat more fragile. A sufficient shielding would reduce the payload of the humans. Science fiction stories assume the use of a hollowed out asteroid as a radiation shield.

5.3. Light Pressure — Sailing

The rocket technology has basic limitations. One possible alternative is to "sail" on the light pressure of the Sun as a launch "vehicle". For photons, the pressure, P, is proportional to the energy density, u, Eq. (2.18) as follows:

$$P = 4\sigma T^4/3c = u/3 \qquad (5.1)$$

For the pressure due to the Sun, with luminosity L (in W) at a distance r is given as:

$$P = (2L)/(4\pi c r^2) \qquad (5.2)$$

The factor of 2 arises because the sail is assumed to be perfectly reflective and normal incidence of the radiation is assumed. Since the gravity of the Sun falls off with radius exactly as the light pressure, the net magnitude of the acceleration is largest at the start of the trip and falls off as the distance increases. The acceleration can be positive for a reflective sail with sufficiently low density, ρ, and thickness, d. If the payload is small, there is no dependence on sail area because larger area means larger light pressure, which scales. The acceleration depends only on the ρ and d of the sail if the payload is negligible as shown in the following equation:

$$d^2r/dt^2 = [2L/(4\pi c \rho d) - GM]/r^2 \qquad (5.3)$$

As with previous work on inverse square central forces, the acceleration can be integrated numerically to yield the time development of the velocity and position. That is what is done in the App "Solar_Sail_App" using the Matlab utility "ode45". For the sail, a density of $1.5 \times 10^3 kg/m^3$ is assumed to represent a thin aluminized mylar. A 1-μm-thick circular sail of 10-km radius has a mass of about 4.7×10^5kg. The results of the App for a specific Slider choice of initial radius and sail thickness are shown in Figure 5.5. For the final velocity, energy considerations are used since at large r the

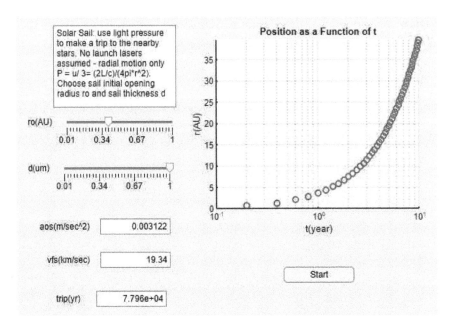

Figure 5.5:   Plot of the sail position for the first 10 years of operation. The initial acceleration, aos, followed by an asymptotic velocity, vfs, is evident. The trip time is for a distance of 5 ly.

"potential" due to gravity and light pressure disappears and the final kinetic energy is used to find the final velocity. The initial acceleration and the final velocity are displayed using EditFields.

In the specific case shown in Figure 5.5, the sail is released at rest from a radius of about 0.4 AU, about the radius of Mercury. The thickness is chosen to be about $1\,\mu$m. The final velocity is about $19\,$km/s and the trip time is about 78,000 years. It is assumed that the sail can be unfolded instantaneously at the starting point and that at the starting point the sail is at rest. The payload needs to be small with respect to the sail mass if the results preceding are to be valid.

These are simplifications, but the basic ideas are valid and the conclusion is that the trip of 5 ly is still a very slow one. Nevertheless, the Sun shines for free. In addition, the technology is probably cheaper and simpler than the other options provide and therefore more probable of being developed. Indeed, small model sails have been deployed within the solar system. A picture of a prototypical sail is shown in Figure 5.6. It seems likely that solar sails will be used as a cheap alternative to rockets, at least for short-distance solar applications.

Figure 5.6: Schematic picture of a NASA sail prototype showing the thin aluminized body of the sail and the ribs used to extend the sail.

5.4. Ion Exhaust

The rocket equation implies that the payload ratio, Eq. (4.18), scales exponentially with the exponent being the final velocity divided by the exhaust velocity. For a large final velocity, that puts an emphasis on increasing the exhaust velocity beyond what is available to conventional rocket fuel such as liquid hydrogen mixed with liquid oxygen with velocities of order $5\,\mathrm{km/s}$. Otherwise, the payload ratio becomes very small and the trip times become too large.

One idea is to accelerate charged particles to a high energy and eject them. Here, only simple electrostatic ion acceleration is assumed. The engine is operated in "space charge limited" mode because the dense ion current distorts the accelerating electric field and reduces it. This effect sets a limit on the ion current density. The ion beam also repels itself since the charges are the same, but this effect of intense beams is ignored here. For the fields and number densities explored here, the beam–beam repulsion effects are small with respect to the overall external accelerating electric field.

Note that the force exerted per unit area, A, or the energy density, u, of the accelerating electric field, E, is

$$F/A \sim \varepsilon_o E^2 / 2 \qquad (5.4)$$

For "plausible" fields of order $100\,\mathrm{MV/m}$, the F/A is about $40{,}000\,\mathrm{Nt/m^2}$ which is about 100 times weaker than that provided by conventional rocketry. This disadvantage can be ameliorated because very large exhaust velocities can be obtained and the engine need not have such a limited burn time as a rocket since the ejected propellant mass, m, is an ion current and is much reduced in density compared to a liquid.

Consider a gas with ions of mass m, and charge q which is accelerated in a gap of length d which exists from z of 0 to $z = d$, with a total potential step of Φ_o. The Poisson equation for the electrostatic potential, Φ, due to an ion current in the accelerating gap for a number density, n, of ions is

$$d^2\Phi/dz^2 = -qn/\varepsilon_o, \quad \Phi(0) = 0, \quad \Phi(d) = -\Phi_o \qquad (5.5)$$

The further boundary condition that the derivative of the potential, the field, at $z = 0$ vanishes means that the regime is space charge limited by the current but that the current does not disturb the static potential.

The constant number density n in the gap defines a current density per unit transverse area, j, of ions. The solution is easily found by assuming that the potential obeys a power law in z with an exponent found by substitution into Eq. (5.5) to be 4/3 since $qn/\varepsilon_o = j/(\varepsilon_o v)$ and $v \sim$ the square root of Φ. The maximum of j, j_o, occurs at the gap exit with an exit velocity of the ions of v_o. The exit velocity follows from energy conservation assuming ion injection into the field region occurs approximately at rest:

$$\Phi(z) = -\Phi_o(z/d)^{4/3}, \quad E(z) = -\partial\Phi/\partial z = (4/3)(\Phi_o/d)(z/d)^{1/3}$$
$$j_o = qnv_o = [(4\varepsilon_o/9)\Phi_o^{3/2}\sqrt{2q/m}]/d^2, \quad v_o = \sqrt{2q\Phi_o/m}$$

$$(5.6)$$

The behavior as a rocket with mass ejected per unit time, dm/dt, assuming a transverse area of A occupied by the accelerating gap, then follows from Eq. (4.12). The value of dm/dt is a constant which depends only on the electric field, the gap area and the exhaust velocity. It will be fixed here to an unrealistic value, which will be discussed in a Section 5.9

$$dm/dt = j_o Am/q = nAmv_o = [8\varepsilon_o(\Phi_o/d)^2 A]/9v_o \qquad (5.7)$$

To optimize the acceleration, Eq. (5.8), in this simple electrostatic device, maximize the electric field. The ion type is not relevant as q and m drop out of the expression for acceleration. The total rocket mass, M, decreases in time as dm/dt which was also the case for a classical rocket. As expected from Eq. (5.4), the force per unit aperture area A depends only on the square of the electric field, i.e.,

$$(d^2r/dt^2) = [A(\Phi_o/d)^2(8\varepsilon_o/9)]/M \qquad (5.8)$$

A schematic of such a device is shown in Figure 5.7. Electrons are injected into a chamber and ionize a propellant gas. The ions are captured by a magnetic field and accelerated through an electric field. To avoid a charge up of the device, a second electron gun neutralizes the ions after acceleration. The magnetic field can be superconducting, so as to reduce the power needs of magnetic field generation.

A simplified look at an ion rocket is made using the App "Ion_exhaust_3". The total initial mass is chosen by the Slider, while the exhaust is driven by singly ionized atoms, with mass chosen by Slider also. The exhaust transverse aperture, the accelerating voltage, and the accelerating gap are all chosen by Slider. The payload ratio is also chosen by Slider. A set of results for a specific choice of Sliders is shown in Figure 5.8. The symbol V is used instead of Φ because

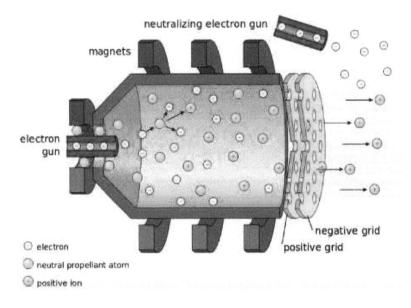

Figure 5.7:　Schematic diagram of an electrostatic ion exhaust rocket engine.

that symbol is not available in the text box for the App nor is ε_o for the symbolic acceleration EditField. The results for v_o in km/s, initial acceleration dv/dt, in m/s^2, dm/dt in kg/s and total burn time T in seconds are displayed using EditFields. For this choice, the acceleration initially is $\sim g$.

The field is approximately 10 MV/cm with a transverse area of about 5 m^2. The initial rocket mass is $10^{6.5}$ kg with a 30% payload in this example. The exhaust velocity is large, about 14% of c, and the acceleration is $\sim g$. The mass rate is about 0.8 kg/s. The burn time is about 1 month at the end of which the rocket has a velocity greater than v_o of about 20% of the velocity of light. Assuming that the velocity is constant at its final value, a trip of 5 ly takes about 30 years. Admittedly, the example is very optimistic, but it does illustrate the virtue of improving the exhaust velocity using particle acceleration. A possible method of interstellar exploration is available for trips of decades to the nearest stars rather than thousands of years. These types of rockets are potentially very useful and NASA has in fact deployed some test models. The technique is considered

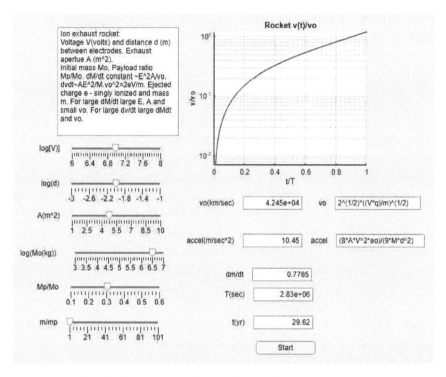

Figure 5.8:   Results for a specific choice of Sliders for operation of the ion exhaust rocket.

very competitive for longer missions, and alternative technologies are presently being field-tested.

5.5. Ramjet

Why bother carrying reaction mass when it is there for the taking? Ramjets are common objects when used in the Earth's atmosphere. The concept for interstellar travel is to scoop up your fuel from the interstellar medium, (ISM). Collect protons which are the most common primordial element in the ISM as shown previously in the discussion of protostars. Somehow, the protons are collected from the ISM using a scoop of transverse area A. Perhaps the hydrogen is ionized and collected by a magnetic scoop. The mass collected increases with the velocity of the ramjet. That is the basic appeal of a ramjet. The density of the ISM is ρ and the ramjet velocity is v so the rate of mass being scooped up is given as

$$dm/dt = A\rho v \qquad (5.9)$$

The ramjet mass is M. Ignore losses due to engine heating or input fuel not participating in the postulated fusion reaction. For a mass dm scooped up, a fraction is converted to exhaust, which is also taken here to be 100% efficient. The scooped up protons supply a fusion reactor with fuels to fuse into helium. The idea is to pick up your fuel on the voyage. Or perhaps one can dip into the upper atmosphere of a planet for high-density refueling.

For low velocities, there is no need for shielding from the energetic collisions with the medium. However, at relativistic velocities, the scoop is colliding with high-energy protons. As shown in Figure 4.15, the collisions would initiate showers of secondary particles which would irradiate the crew. This is a serious problem for a ramjet and requires some method of shielding just as it is for the ion rocket and any craft going at a substantial fraction of c with respect to the ISM. Indeed, magnetic shielding as postulated by science fiction writers does not work on the neutral hydrogen in the ISM so the ramjet has to ionize the hydrogen and then sweep it aside.

Assume the ramjet carries a neutron "bottle" to start the fusion reactions and a method to ionize the H molecules if the ISM is locally neutral. A possible chain of fusion reactions is given as follows:

$$n + p \rightarrow D + 2.2\,\text{MeV}$$
$$D + D \rightarrow T + p + 4\,\text{MeV} \qquad (5.10)$$
$$D + T \rightarrow H_e^4 + n + 17\,\text{MeV}$$

You get back some neutrons at the end of the cycle which can be reused but would not be contained by the magnetic fields which confine the plasma. The other "cinders" are helium nuclei.

The ramjet needs an initial boost, v_i, to get started since it must scoop up the fuel in order to accelerate. Assume an initial velocity of 30 km/s comparable to velocities seen in solar exploration and provided by disposable rockets or some other means. The ramjet speed increases due to the ejection of helium of mass dm with a speed roughly $0.05\ c$ or v_o with respect to the rocket. The velocities are not relativistic because the binding energies are of scale MeV as in Eq. (5.10), while masses are of scale GeV.

An artists' view of a fusion ramjet is shown in Figure 5.9. The scoop funnels the ISM down into a fusion reactor which then ejects the energetic charged helium nuclei. Given the size of the ITER device, the optimism seems excessive. In any case, think "Back to the Future" and assume a slimmer fusion reactor will appear in the future. More discussion on energy conservation will appear at the end of this chapter.

Figure 5.9: Artists' conception of a ramjet using ISM fuel to power a fusion reactor.

A range of known ISM compositions is shown in Figure 5.10. The protons can exist as molecular hydrogen or as ionized protons and electrons. The density varies dramatically with a low value $\sim 10^6/\mathrm{m}^3$. Really efficient ramjet operation implies working with a rich fuel of high density as exists in molecular clouds. The local galaxy, the Milky

Component	Fractional volume	Scale height (pc)	Temperature (K)	Density (particles/cm^3)	State of hydrogen
Molecular clouds	< 1%	80	10–20	10^2–10^6	molecular
Cold Neutral Medium (CNM)	1–5%	100–300	50–100	20–50	neutral atomic
Warm Neutral Medium (WNM)	10–20%	300–400	6000–10000	0.2–0.5	neutral atomic
Warm Ionized Medium (WIM)	20–50%	1000	8000	0.2–0.5	ionized

Figure 5.10: Table of a range of ISM that might be encountered (from Wikipedia).

Way, has a mean ISM number density which varies tremendously by a factor $\sim 10^{-7}$ with position.

The acceleration of the ramjet, with mass M, is proportional to velocity which means the ramjet velocity increases exponentially. The energy yield of the fusion reactions is Q. The characteristic exponential time scale is T_r. The appeal of scooping up fuel on the way is that payload fraction is not an issue and that a steadily increasing acceleration yields an exponential increase in velocity.

$$dv/dt = [(\rho A v)/M]v_o, \quad v_o/c = \sqrt{2Q/(m_{\text{He}}c^2)}$$
$$T_r = (M/\rho A v_o) \tag{5.11}$$
$$v(t) = v_i e^{t/T_r}, \quad r(t) = [v(t) - v_i]T_r$$

The behavior of a ramjet is studied using the App "ramjet_APP". The Q value is fixed at 5 MeV. A scoop of $1\,\text{km} \times 1\,\text{km}$ is assumed. The local ISM number density of protons is chosen via the Slider as is the ramjet mass. The exponential time scale, v/c, at the end of a century and the distance attained are shown using EditField outputs. The result of a specific set of input parameters is shown in Figure 5.11. With the optimistically assumed values of the parameters, a shorter trip over distances of light years appears to be possible. All that is needed is a compact and light fusion reactor to heat, compress and eject the scooped up fuel.

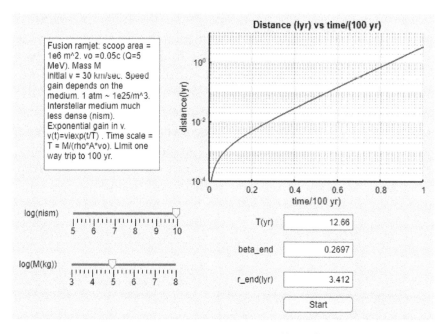

Figure 5.11: Ramjet behavior for the choice of $10^{10}p/m^3$ in the local ISM and a total ramjet mass of 10^5kg. A trip of 3.4 ly is then possible in a century.

Since both the ion exhaust rocket and the fusion ramjet could attain final velocities which are a substantial fraction of the speed of light, the limitations due to SR need to be addressed. Relativity has some very interesting modifications to the issues of interstellar travel.

5.6. Time Dilation

So far, interstellar travel looks difficult. It seems clear that higher velocities will be needed to make a trip to extra-solar planets within the lifetime of a person. Conventional rockets or solar sails seem to require trips of tens of thousands of years. Use of ion exhaust or fusion ramjets might reduce the trip to the nearest stars, but the technology is still rather far off. It is time to see if relativistic velocities can be obtained. If so, the nearest stars would only require trips on a scale of years. However, if the velocities are a substantial fraction of the speed of light, the physics will need modification.

Einstein showed that time depends on the frame of reference because the speed of light is the same in all reference systems moving in uniform relative motion, or inertial frames. Time dilation was already introduced, albeit somewhat abstractly, in Eq. (3.21) by invoking the invariant interval between two space–time events, ds. Time on moving clocks was shown to be greater than time on clocks at rest. At this point, a very simple proof is made of the same result by constructing a "clock" using light and simply invoking geometry as seen in Eq. (5.12) and Figure 5.12 which was created by the App "SR_Clock". A movie of the light path in both frames is made and compared.

$$t_* = (2L/c), \quad t = \gamma t_* = t_*/\sqrt{1 - \beta^2}, \quad \beta = v/c \qquad (5.12)$$

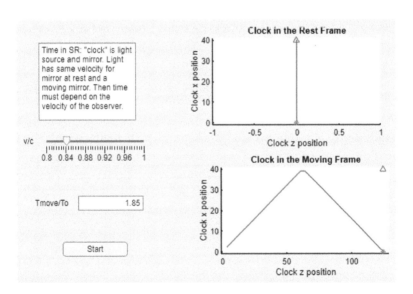

Figure 5.12: Construction of time in terms of "clock ticks" in a proper frame with observers at rest and time as seen in a frame where the clock moves. In both cases, the light velocity is c.

The "clock" is a light source and a mirror. At rest, the clock ticks in a time $t_* = (2L)/c$. In the frame where the clock has a velocity v, the light travels further, but with the same velocity of c, so the time is dilated. The horizontal half distance is $vt/2$ and the vertical height is $L = ct_*/2$, and distance transverse to the motion is invariant. The triangle has a hypotenuse of length ct. Since β is less than 1 by hypothesis, t is greater than t_* and the time seen by observers in a frame where the clock is moving is greater than the time for observers at rest with respect to the clock.

5.7. Relativistic Rocket — g

The previously considered devices began to attain exhaust velocities near that of light in the case of ion exhaust rockets and ramjets. Because of this, SR physics must be used. However, all that known physics allows is some form of momentum conservation, so that "rocketry" is the only tool available at present. Before looking into those aspects, first consider a rocket somehow moving with a constant proper acceleration of g. By "proper" is meant what the passengers would experience. Classically, $v = gt$, and in 1 year a rocket with constant acceleration of g reaches $v \sim c$.

The Lorentz transformation equations relating position and time in two reference frames in relative motion are shown in Eq. (5.13). The spatial and temporal origins coincide at time zero. The relative velocity of the two frames is β_o as can be seen by setting $z_* = 0$. If the $*$ frame is proper, $z_* = $ constant and $dt = \gamma_o dt_*$ and time dilation is recovered. Velocity addition is derived by differentiating the position and the time in the two frames. If $\beta_* = 1$, then $\beta = 1$ and light has the same velocity in all frames. If $\beta_* = 0$, then $\beta = \beta_o$ by definition. If the velocities are small, the classical addition of velocities is recovered.

Differentiating the velocities, it can be shown in the special case where one frame is proper, here the $*$ frame, that $\gamma^3(d\beta/dt) = \gamma^3 a = a_*$, where γ and β refer to the relative velocity between the two reference frames with accelerations, and a_* is the proper acceleration. The observer in the other frame, since the γ factor increases under acceleration, sees a reduced change in velocity as the velocity of the ship approaches c since c cannot be exceeded.

$$z = \gamma_o[z_* + \beta_o(ct_*)], \quad ct = \gamma_o[(ct_*) + \beta_o z_*]$$
$$\beta = dz/d(ct) = (\beta_o + \beta_*)/(1 + \beta_o\beta_*) \tag{5.13}$$
$$a = d\beta/cdt = a_*/\gamma_o^3, \quad \gamma_o = \gamma, \quad \beta_o = \beta$$

The equation for a can be integrated using $t = \gamma t_*$, and β as a function of t_* appears in Eq. (5.14). Then γ can be calculated. Using the expression for γ to relate dt and dt_*, the expression for t follows. Another integration using $dz = \gamma\beta(cdt_*)$ gives the position z as a function of t_*. The parameter α has dimension of inverse time while q is dimensionless. The passengers feel a constant and comfortable acceleration, $g = a_*$.

$$\beta = \tanh(q), \quad \gamma = \cosh(q), \quad \alpha = g/c, \quad q = \alpha t_*$$
$$t = \sinh(q)/\alpha, \quad z = c[\cosh(q) - 1]/\alpha$$

$$(5.14)$$

These equations are explored using the App "Rocket_SR_g" where a specific output figure of the App is shown in Figure 5.13.

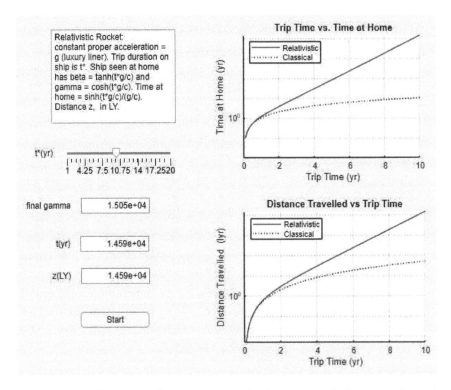

Figure 5.13: Classical and relativistic results for times and distances of travel at a constant proper acceleration of g. The time elapsed at home is t while it is always t_s classically. For the first year of travel, the two curves on the plots approximately agree.

The only choice is the Slider for the duration of the trip. The classical results are also shown for comparison. The vertical axes are logarithmic, so the differences from the classical results become profound for longer trips. For a 1-year trip, the time at home is only 1.2 years. However, after 10 years of constant proper acceleration, the time is about 32,000 years and z in light years is approximately t in years. A 20-year trip spans 440 million ly for the stay-at-home observer.

In the specific case shown the times are about a thousand times different and the distances more than a hundred times different, comparing the classical and relativistic results. For very short trips, the classical results, for example, $v = gt$, $t = t_*$ are recovered. If such a rocket could be constructed, passengers could visit the galaxy in trips of a few decades. However, they would forever lose contact with home because many relativistic velocities would be employed. Since the known physics available to travelers is only momentum–energy conservation, a realistic relativistic rocket needs to be addressed to see if such travel is possible.

5.8. Relativistic Rocket — Fuel

The question is now can such a rocket, or one approximately like it, be constructed for the passengers to avail the mixed benefits of relativity, shorter trip times, and the loss of home contacts? The classical rocket equation first needs to be reformulated in SR. There are some algebraic results that are useful; the differentials $d\gamma = \beta\gamma^3 d\beta$ and $d(\beta\gamma) = \gamma^3 d\beta$ which were previously used in the derivation of the proper acceleration, Eq. (5.13). These differentials are easily established using Matlab symbolic math utilities. In general, the user need almost never need to find an integral, derivative, or solution to an ordinary equation or of a differential equation when the symbolic tools are available. Code snippets are shown in Figure 5.14 which find the differentials.

As before in the classical case, consider the "decay" of a rest mass m moving with velocity βc into a mass $m - dm$ moving with velocity change $d\beta$ and an expelled mass moving with velocity $\beta_o c$ with respect to mass m in the rocket rest frame. In SR, mass is

```
>> syms p b
>> p = 1/sqrt(1-b^2);
>> pretty(simplify(dp))
      b
   -----------
       2 3/2
   (1 - b )

>> p = b/sqrt(1-b^2);
>> dp = diff(p,b);
>> pretty(simplify(dp))
      1
   -----------
       2 3/2
   (1 - b )
```

Figure 5.14:   Code snippets to find the derivatives of γ and $\beta\gamma$ to be $\beta\gamma^3$ and γ^3, respectively.

not conserved and the m used here is the mass of an object at rest. The momentum is also changed. Energy $\varepsilon = \gamma mc^2$ and momentum $p = \gamma m\beta c$ in SR and appear in Eq. (5.15), where c is set $= 1$ for convenience. The NR limit for ε is $mc^2 + mv^2/2$, while for momentum it is mv. Energy and momentum are conserved and are treated together as (ε, pc) with the same dimensions as was the case for (z, ct).

A simple example is a charge, q, in a uniform electric field, E. Classically, the force is qE and the velocity increases as $v = (qE)t$ without limit. In SR, it is the time rate change of momentum that is equal to qE, or $p = (qE)t$. The velocity is $\beta = cp/\varepsilon = (p/mc)/\sqrt{1 + (p/mc)^2}$, which has an upper limit of 1, as expected.

Energy conservation and momentum conservation are written in the frame where the rocket has velocity βc. The velocity of the fuel in that the frame is related to the fuel exhaust velocity in the rocket rest frame using SR velocity addition, already derived in Eq. (5.13).

The mass, velocity, and energy of the fuel have the subscript f, i.e.,

$$\varepsilon = \gamma m, \quad p = \beta \gamma m$$

$$d(m\gamma) = -\gamma_f dm_f, \quad d(m\beta\gamma) = \beta_f \gamma_f dm_f$$

$$\beta_f = (\beta_o - \beta)/(1 - \beta_o \beta) \tag{5.15}$$

$$d(m\beta\gamma) = [(\beta - \beta_o)/(1 - \beta_o \beta)]d(m\gamma)$$

Using the differentials $d\gamma$ and $d(\beta\gamma)$ quoted above, after some tedious algebra the SR differential equation for the rocket is shown in Eq. (5.16). The first integral can be done in closed form. Compared to the classical result, $dm/m = -dv/v_o$, as β approaches 1, the effect of the exhaust velocity on the rocket velocity is reduced by the factor γ^2, as it must be since β is always less than 1 for material objects. The classical limit is obtained when β is small, as expected, i.e.,

$$dm/m = -d\beta(\gamma^2/\beta_o)$$

$$m_f/m_i = [(1 + \beta_i)/(1 - \beta_f)]^{1/(2\beta_o)} \tag{5.16}$$

$$\beta_f - \beta_i = (m_f^{2\beta_o} - m_i^{2\beta_o})/(m_f^{2\beta_o} + m_i^{2\beta_o})$$

The consequences of a correct rocket description for relativistic motion are explored in the App "SR_Rocket_a". The equation for $d\beta/dm$ is solved symbolically with the initial condition $\beta(0) = 0$ using the Matlab utility "dsolve", and the integration of β to evaluate the distance is accomplished numerically using the Matlab utility "quad" by defining a constant burn rate to convert m into time. This rate can be changed by using the Slider. The inverse equation for $dm/d\beta$ is also solved symbolically with the initial condition $m(0) = m_i$. The proper acceleration is evaluated using Eq. (5.13), $a_* = a\gamma^3$, and a is found using $(d\beta/dm)(dm/dt)$ with a constant dm/dt assuming that the fuel is exhausted in a time T set by the defined payload fraction. The maximum and minimum proper acceleration occurs at the start and the end of the burn. If the proper acceleration is to be kept $\sim 1g$ or less, the exhaust velocity should be kept low or the burn time must be extended.

Mass/energy is expelled and it could be photons since they carry both energy and momentum. Photons would clearly give the greatest possible exhaust velocity, one much greater than chemical rockets or even ion exhaust rockets. All one needs do is create antimatter in macroscopic quantities on board and annihilate it into photons or perhaps accelerate charges, electrons, bend them, and thus radiate energetic bremsstrahlung photons with energies ~keV, aimed in a specific direction. Less exotic methods might also be used.

One specific set of parameters is shown in Figure 5.15. A constant burn rate dm/dt is set to keep the proper acceleration at a

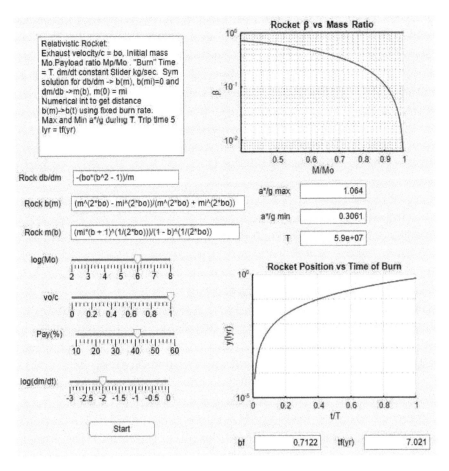

Figure 5.15: Results for a burn of a relativistic rocket which has exhaust velocity, initial mass, mass burn rate, and payload percentage chosen by the Slider. The final velocity and trip time are displayed as well as the proper accelerations and the total burn time.

comfortable value. In this case, the final β factor is about 0.72 and a 5-ly trip takes 7 years. The exhaust velocity, the horizontal red line in the plot of rocket velocity, is taken to be that of photons. To limit the acceleration, the mass loss rate is set at $dm/dt - 0.01\,\mathrm{kg/s}$. The mass is 10^6 kg and a 40% payload is chosen. The "burn" time is about 2 years.

In Figure 5.8, for ion exhaust the exhaust velocity was $\beta_o = 0.14$ and with a 30% payload the travel time was ~ 30 years for a 5 y trip with acceleration less than g and $dm/dt \sim 0.8\,\mathrm{kg/s}$ using protons. The Sliders can be used here to approximately crosscheck the ion exhaust results. Setting the Sliders for the case where $M_o = 10^{6.5}$ kg, $\beta_o = 0.14$ and a 30% payload fraction, with $dm/dt = 1\,\mathrm{kg/s}$, the results are $\beta_f = 0.17$, $a_* = 4.6\,g$, $T = 2.3 \times 10^6$ s and the trip time is 29 years which checks reasonably well. The effects of SR are dramatic only when the γ factor is $\gg 1$.

The time disparity depends on γ which is ~ 1.09 and not β, which means the passengers would not age very differentially compared to the home folks. Therefore, a trip of 5 ly in a reasonable amount of time is at least theoretically feasible. In fact, the need for shielding is much reduced since the ISM the rocket collides with is much less energetic in this case since the energy which the rocket interacts with ISM also scales with γ. If the rocket can use photons for thrust, with a 40% payload ratio the final β factor is 0.71 which shortens the trip considerably but makes shielding and time slippage more of a problem.

5.9. Energy and Power Issues

Typical chemical rockets, Figure 4.7, have $v_o \sim 3\,\mathrm{km/s}$, dm/dt 1.5×10^4 kg/s, and burn time $T \sim 270$ s. For an ark rocket, the energy for exhaust is supplied by chemical reaction of the fuel and the oxidizer. For example, dynamite has an energy yield of 5.6×10^6 J/kg. Even with small payloads, a trip of 5 ly takes thousands of years. For a solar sail, there is no fuel to carry, but the energy transfer to the sail from the Sun is very small, and the trip also takes a long time.

There are energy and power issues with the ramjet because the ISM must be collected, compressed, heated, and ejected as a focused beam. The details are complex but the issues of just supplying the energy, but not the ISM fuel, are still formidable. In the case of a photon exhaust rocket, the energy issues are comparable to those of the ion exhaust rocket in severity. Since the ion rocket is quite specific in design, it will now be studied in regard to energy and power issues.

Consider the energy and power requirements of the ion exhaust rocket in a bit of detail as a simple example. For the ion exhaust rocket, the motion is almost NR, so classical physics can be used. The fuel to be expelled is carried on the vessel and the acceleration of the ion proton beam must use on-board energy supplied to maintain the electric field in analogy to the oxidizer and plumbing of a conventional rocket. The energy needed for the magnetic field and the energy needed to ionize the assumed LH2 fuel are ignored.

Possible on-board sources of that energy are a fission reactor or a fusion reactor. Using fission, Section 2.11, the energy yield is 8×10^{13} J/kg. For D–D fusion the yield is $\sim 3 \times 10^{14}$ J/kg, which is $\sim 10^8$ times more energy per mass than chemical processes. These yields are much larger, but the ejected mass is an ion beam, with a small density. A fusion reactor will be assumed to exist as a power source for the rocket.

Assume that protons are exhausted. Their initial kinetic energy, K_p, is eV. Assuming $V = 10^7$ volts, the K_p value is 1.6×10^{-12} J/p and the velocity is 4.4×10^7 m/s. Assume the fusion reactor produces 10^{11} W. This is about 100 times larger power output than the terrestrial fission reactors now in place. Protons are exhausted at a rate of 6.2×10^{22} p/s or $dm/dt = 1.04 \times 10^{-4}$ kg/s, which is $\sim 10{,}000$ times smaller than what was previously assumed. By keeping v_o constant, reducing the aperture area by a factor 10 and increasing the voltage grid to 0.32 m from 0.01 m, this reduced value of dm/dt can be provided by the postulated fusion reactor.

Requiring the final velocity to be the same as before, $\beta_f = 0.147$, the ratio of final to initial mass is ~ 0.37. For a total burn time of 1 year, the reactor must produce 3.15×10^{18} J, which could be supplied by $\sim 10^4$ kg of D–D fuel. Assuming a 1% efficiency of fuel mass to "plumbing" for containment, shielding, power generation, and other overheads, there is about 10^6 kg of "plumbing" in the final mass. For a useful payload of 10^6 kg, the plumbing is about half again, so that the initial mass would be $\sim 6 \times 10^6 kg$. The acceleration at the end of the burn would be 2.3×10^{-4} g, much reduced because dm/dt was reduced to a level that could be supported by the total power of the fusion reactor.

Comparing these numbers to those shown in Fig. 5.8, the voltage is the same and so is the exhaust velocity. The gap is increased and the aperture decreased to map into a 100 GW postulated fusion reactor. The total mass and payload mass fractions are similar, except the mass of the fusion reactor is taken into account now. The acceleration is decreased by a factor of a few thousand, as is the dm/dt value. This scenario is not theoretically impossible, although it assumes a small and light fusion reactor.

5.10. Other Possibilities

"Human uploads have such a natural advantage over present-day people in the environment of space. It's exceedingly unlikely flesh-and-blood beings will ever engage in interstellar travel."

— **Frank Tipler**

"I became interested in this question of whether you can build wormholes for interstellar travel. I realized that if you had a wormhole, the theory of general relativity by itself would permit you to go backward in time."

— **Kip Thorne**

So far the emphasis in this chapter has been on individuals traveling to the stars in a finite period of time. Perhaps easier tactics should be contemplated given the apparent difficulties of this path. It should be possible to transport frozen embryos instead of adults and then birth them at the end of the mission, thus reducing the payload requirements. The important thing is information transfer, so one can imagine sending only data and using three-dimensional printers at the end of the trip which would use the materials that are available at the destination. The idea would be to send information not hardware or actual people.

The ultimate human information is the genetic code. Humans have 23 chromosomes. There are four bases and four possible base pairs within the structure of DNA. With two parents, there are 6×10^9 base pairs to define an individual. For storage, 1 byte is 8 binary bits and can represent 4 base pairs. Overall, the genome information could be contained in a file of about 15GB. That is less than the capacity of a thumb drive. If data degradation during the trip, due to radiation upsets or other causes, can be avoided by fault-tolerant or redundant information storage, then the information is still quite compact. In fact, if there were a functioning receiver, the information might be sent at the speed of light and the code used to establish humans at the receiving location.

What about actual people and not their genomes? The number of cells in a human body is estimated to be about 100 trillion. Imagine scanning a person instantaneously, which given the finite speed of light is a wildly optimistic assumption. Unfortunately, Avogadro's number is still an issue. For an object containing $1m^3$ of water, you need to scan, record, transmit, and receive the vector position and velocity of about 10^{29} H_2O molecules. Clearly, "transporters" are right out. There is just too much information to extract, send, and then receive, not to mention that fault-tolerant procedures are mandatory.

As for "warps", "wormholes", and other science fiction-postulated devices, this text restricts itself to physics that is presently understood. Regrettably, the speed of light appears to be a limiting velocity because SR has been tested innumerable times and found to be valid. The conservation of energy and momentum are also well tested and the conclusion is embodied in the kinematics of the relativistic rocket which was explored above. Within that limited context, trips to the nearer stars are possible with a finite duration.

5.11. Conclusions

"Space doesn't offer an escape from Earth's problems. And even with nuclear fuel, the transit time to nearby stars exceeds a human lifetime. Interstellar travel is therefore, in my view, an enterprise for post-humans, evolved from our species not via natural selection, but by design."

— **Martin Rees**

"If interstellar travel is as time- or energy-demanding as the above figures indicate, it is far from obvious what the motive for colonization might be."

— **Barney Oliver**

"We must cultivate our own garden."

— **Voltaire**

Indeed, human beings have evolved to fit into the Earth's environment with food, water, gravity, and cosmic ray protection. The idea of a Garden of Eden is a response perhaps to a recognition of the niche that humans fill on the Earth. All the things for survival need to be available during solar system missions in addition to propulsion and communications. For interstellar exploration, the issues are highly complex due to the much vaster distances involved. Perhaps, humans should first secure and cultivate the Earth as a garden spot before boldly going forth. In any case, interstellar travel of any sort will require radical improvements in technology which will not become available in near future.

Chapter 6

Additional Miscellaneous Topics

"Computers are like Old Testament gods; lots of rules and no mercy."

— **Joseph Campbell**

"The purpose of computing is insight, not numbers."

— **Richard Hamming**

6.1. EM–FFT for Poisson Equation

The last chapter of this book takes a brief look at other tools that are available in Matlab that can be applied to problems in other areas of physics. Matlab is a very extensive suite of tools, a suite which continues to grow. It can be used to explore many different problems. The first example is Matlab applied to solving boundary-value problems (BVP) numerically as opposed to the BVP problems addressed previously using specialized Matlab utilities.

The Poisson equation is the differential equation defining the electrostatic potential, Φ, due to a distribution of charge density, ρ. It was already introduced in Eq. (5.5) in the problem of the ion exhaust rocket, where a closed, form solution in one dimension could be found. For problems that are only tractable numerically, the equation in two dimensions can be approximated on a grid of $N \times N$ points (x_i, y_j) with grid spacing δ as seen in Eq. (6.1). The grid can be made as fine as is necessary to approximate the charge distribution to the desired accuracy, i.e.,

$$\nabla^2 \Phi = -\rho/\varepsilon_o$$
$$4\Phi_{i,j} - \Phi_{i+1,j} - \Phi_{i-1,j} - \Phi_{i,j+1} - \Phi_{i,j-1} = \rho_{i,j}\delta^2/\varepsilon_o \qquad (6.1)$$
$$\tilde{\Phi}_{i,j} = (\rho_{i,j}\delta^2/2)/[\cos(2\pi i/N) + \cos(2\pi j/N) - 2]$$

The solution is found using the grid to approximate the differential equation by transforming into the frequency domain using fast Fourier transforms with the Matlab utility "fft2", solving algebraically, as shown in Eq. (6.1), and transforming back using the inverse transformation, "ifft2". The solution for the transformed potential is a simple algebraic expression.

The electric field is derived from the potential, Φ, using the Matlab utility "gradient" and plotted using the tool "quiver". The output of the App "EM_Poisson_Rect_App" for a particular choice of charge density is shown in Figure 6.1. The user can change the extent of the rectangular charge using Sliders to pick the location of all four rectangular boundaries independently. The grid boundary is (0,1) in both x and y. The grid has 50×50 points.

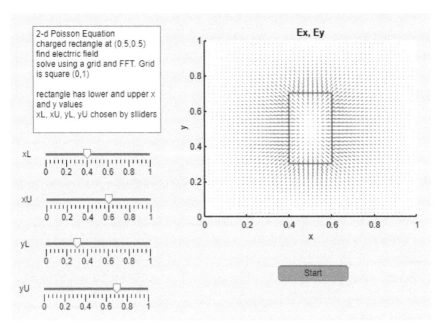

Figure 6.1: The electric field derived from the potential caused by the uniform rectangular distribution of charge which is chosen using the Sliders to set the four boundaries of the charge distribution. The field is defined to vanish on the grid boundaries, so they must be far from the charge distribution.

The field is small inside the rectangle and then falls off outside the charge distribution. If the user is interested, the code can be easily altered to define more grid points and the changes to the field can be tracked since the result is only a numerical approximation to the problem. The potential is assumed to vanish on the grid boundaries, as seen in Eq. (6.1) for the transformed potential. Therefore, the user must position the charges sufficiently far from the border so as not to distort the solution since the correct boundary conditions are only these where the potential vanishes at very large distances from the source charge distribution.

6.2. Quantum Scattering Off a Barrier

There are utilities in Matlab to solve partial differential equations (PDE). An example is given here to solve the one-dimensional Schrödinger equation, Eq. (6.2), familiar from non-relativistic quantum mechanics. A useful way to remember the equation is to think of it as the quantum operator version of $U = K + V$ operating on the wave function ψ. The energy is proportional to the time rate of change of the wave function while the momentum is proportional to the spatial rate of change and $K = p^2/2m$. For a plane wave, Eq. (2.19) is for photons and Eq. (3.4) is for electrons

$$i\hbar \partial \psi/\partial t = -(\hbar^2/2m)\partial^2 \psi/\partial x^2 + V(x)\psi$$
$$\varepsilon = \hbar\omega, \quad p = \hbar k$$

(6.2)

The system is described by ψ, which maps the evolution of the system under the influence of a potential $V(x)$. The Matlab utility "pdepe" is used to solve the Schrödinger equation in one dimension for an initial localized wave packet, which is the analog of a classical particle. The wave packet has an initial finite range of values of both position, dx_o, and momentum, $\sim 1/dx_o$, as required by the Heisenberg uncertainty principle. The real part of ψ is initially a Gaussian with central location x_o and standard deviation $\sim dx_o$. The wave packet will spread in x as the wave function evolves in time because of the required finite spread of momentum values.

The units used are those appropriate to atomic systems. Energy, ε, is in electron volts, $1\,eV = 1.6 \times 10^{-19}$ J. Length, x, is in Angstrom, $1\,\text{Å} = 10^{-10}$m $= 0.1$ nm. An Angstrom is a unit appropriate for atomic systems, just as the Fermi $= 10^{-15}$ m is appropriate for nuclei. A reasonable atomic scale for time is the femtosecond ($=10^{-15}$ s). The units of $\hbar$ are energy $*$ time or momentum $*$ length, while $\hbar c$ has the dimension of energy $*$ length

$$\hbar c = 2000\,eV * A, \quad \hbar = 6.6 \times 10^{-16} eV^* \sec$$
$$\Delta x \Delta p \geq \hbar/2, \quad \Delta t \Delta \varepsilon \geq \hbar/2 \qquad (6.3)$$
$$m_e c^2 = 0.511\,\text{MeV}$$

The utility "pdpe" requires functions to define the differential equation, Eq. (6.2), the initial conditions, in this case the initial wave packet, and the boundary conditions, in this case the wave function is required to vanish at the grid boundary in x. The grid of points in x, $(-20, 20)$ Å here, and t, $(0, 2 \times 10^{-15}$ s$)$, is used in this App.

The potential is a constant, but the script could easily be adjusted to handle $V(x)$. The user can choose the depth or height of the potential and its spatial extent using Sliders. A "movie" of the solution is then generated for that choice. A specific frame of the movie for a specific choice is shown in Figure 6.2. The App used is called "QM_Barrier_Scatt_App".

In Figure 6.2, the barrier is about 6 Å wide. The user can explore how the solutions change with the parameters, both the extent and the value of the potential. In quantum mechanics, even the existence of a potential less than the energy of 5 eV causes reflections of the wave packet. Conversely, even for barriers with V_o greater than 5 eV, some fraction of the wave function penetrates the barrier, especially if it is thin. This type of behavior was already discussed in the context of fusion processes where Coulomb barrier penetration was seen to be crucial. The user is strongly encouraged to spend the time to look at solutions covering the full range of possibilities offered by the Sliders.

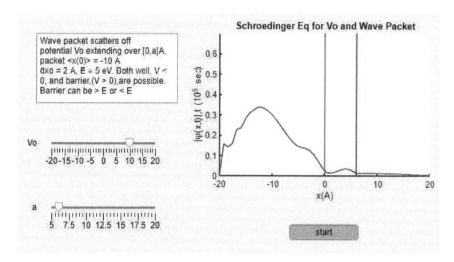

Figure 6.2: Last frame of the movie generated in the scattering of a wave packet off a user-defined constant potential. Both wells, $V_o < 0$, and barriers, $V_o > 0$, can be chosen. The edges of the potential are indicated by vertical red lines.

6.3. Wave Packet in a Well

A variation on the scattering of a wave packet is to consider the behavior of a "particle" bound in a potential well. In quantum mechanics, these classical concepts are only approximate. The wave function generally has a spread in both position and momentum values. Recall the quantum "tunneling through", which enables p–p fusion to occur at temperatures well below those needed classically to overcome the p–p Coulomb repulsion.

The App "QM_Well_Bound_App" explores the behavior of a wave packet localized to a constant potential, taken here to be 0. In this App, a Slider provides the potential outside the well. The user can see what effect the deepening of the well has on the "leakage" of the wave function outside the well and watch a "movie" of the evolution of the wave function. For the specific case where the electron energy is 5 eV and the well is 10 eV deep, the last frame of the movie is shown in Figure 6.3. The oscillatory behavior of a quantum bound state is

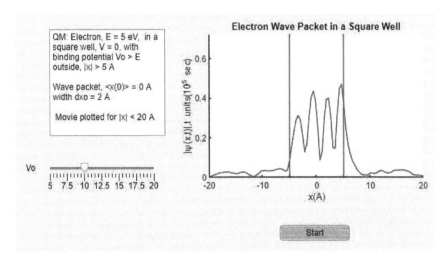

Figure 6.3: Frame of the movie of the time evolution of the wave function bound in a well, where $V_o - E = 5\,\text{eV}$. The edges of the potential are indicated by vertical red lines.

very evident. Again, the user is strongly encouraged to explore the solutions for the range of values available using the Slider. If more freedom is desired, the user can easily edit the App "Code View". Many "comments" are available for the user to understand the code, as is generally true for all the Apps that occur in this book.

6.4. Whale Song

Matlab has utilities to explore sounds in both the time and frequency domains. The utility "sound" was already used to make the "chirp" sound in the App shown in Figure 3.12. A sound file exists in Matlab which has recorded a whale "song" that is played in the App "Whale_App" using the utility "audioread" to read in an audio file much like "dlmread" is used to read in data files in Matlab. The App makes a plot of the amplitude of the sound with the "B Call" selected as a function of time shown in Figure 6.4 and plays it to the speakers using the "sound" utility. The frequency spectrum of this call is derived using the "fft" utility. Since whales vocalize down to 30 Hz and humans can only hear above about 100 Hz, the sound

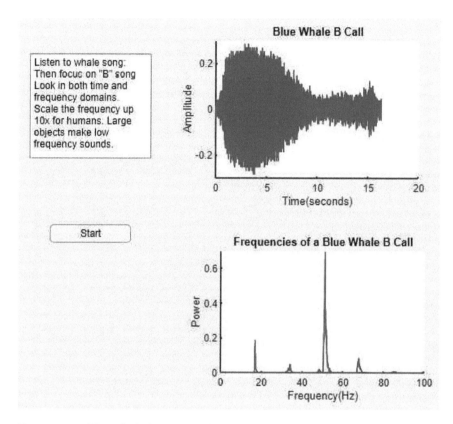

Figure 6.4:   Plot of whale song in the time and frequency domains. The power spectrum shows a few major discrete frequencies.

frequencies are scaled up by a factor of 10 for the "sound" input. Indeed, the maximum power in the B call occurs at about 55 Hz.

6.5. "Pure" Tones

In order to explore this new Matlab ability, the author downloaded a free audio app called "Audacity". Many other free possibilities are available in app stores. The "Audacity" desktop screen is shown in Figure 6.5.

Using that app, the author produced two audio files, .wav format, where the "singing" of a pure tone was attempted. The App "DRG_Tones_App" used the utility "audioread" to load the .wav

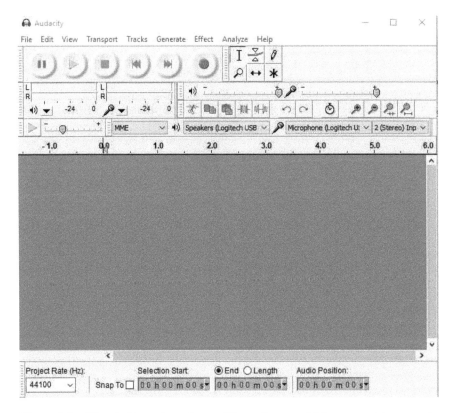

Figure 6.5: The desktop view of a specific audio app which was used to generate .wav audio files.

files and the utility "sound" to play them back on the speakers/headphones. The amplitude as a function of time was then converted to the equivalent description of power as a function of frequency using "fft". The results of the App are shown in Figure 6.6. The "low" file shows several higher harmonics beyond the ~100 Hz main tone. That frequency is near the low-end range of human auditory sensitivity. The "high" tone has significant power at about 350 Hz as seen in the lower plot. There are approximately 7 periods in the upper plot covering 0.02 s or a frequency of 350 Hz which shows the complementarity of the time and frequency domains. The user is encouraged to explore other, more interesting, sounds since Matlab provides the tools to do so.

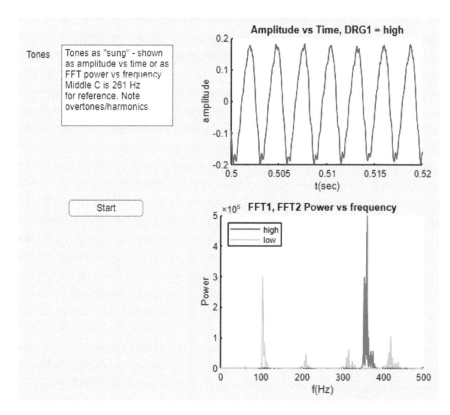

Figure 6.6: Figure generated by the App "DRG_Tones_App". The plot in the time domain shows that the "high" tone file is a reasonable approximation to a sinusoid. The plot in the frequency domain shows the low tone with three visible harmonics and the high tone above middle C.

6.6. Vacuum Energy — Horizons in de Sitter Space

The Apps in this text have utilized only a very restricted use of the objects available to use in the Matlab App suite of tools shown in the "Design View". There are drop-down menus, radio buttons, knobs, and switches to define input parameters. There are also many ways to display results such as tables and different plotting options. The user should not take such a constricted view of the plethora of available tools as the author has but explored the full range of Matlab tools.

In this section, the use of "gauges" is made instead of the "Edit Field numerical" which has normally been the method used to display numerical results. This App, the "deSitterHoriz_App", is also the only cosmological topic raised in the entire book since the earliest timeline assumed an almost uniform plasma of protons, neutrons, electrons, and neutrinos as appropriate to the time after the CMB "decoupled". This plasma then evolved into a neutral gas of proton and helium nuclei and weakly interacting neutrinos, which collapsed gravitationally into protostars.

A de Sitter space is a possible solution to the GR field equations. It is one which is rapidly expanding, so much so that a horizon may exist beyond which an observer cannot connect to events even using light signals. The de Sitter space has a constant Hubble parameter, H, such as would occur in a Universe dominated by a cosmological constant, which perhaps is related to the recently discovered dark energy. Normal Hubble parameters due to normal matter and energy decrease as the Universe expands. The scale factor for distance, $a(t)$, in de Sitter space in contrast increases exponentially with time as $a(t) = e^{Ht}$. This behavior implies that a distinct horizon exists.

Light is emitted at time t_e, where all times are in units of H^{-1}, at radius r_e and received at $r = 0$ and time t_r if it is inside the horizon set by t_h. Distances are in r/c units of time and normalized appropriately as times. During the transit time of the light, the space is expanding and it may be unable to reach $r = 0$. The reception time, in H^{-1} units, and the horizon for a given emission radius are given as follows:

$$t_r = -\ln(e^{-t_e} - e^{-t_h}), \quad t_h = \ln(1/r_e) \qquad (6.4)$$

The user can choose the radius of light emission and the time of emission using the Sliders. The reception time and the limiting horizon time are shown on gauges, and the reception time is also shown in more detail using a numeric EditField. Light travels with coordinate velocity c. A short movie is made showing that light travels as a series of equal time green stars. It can be seen that the light appears to slow down, which occurs because the space itself

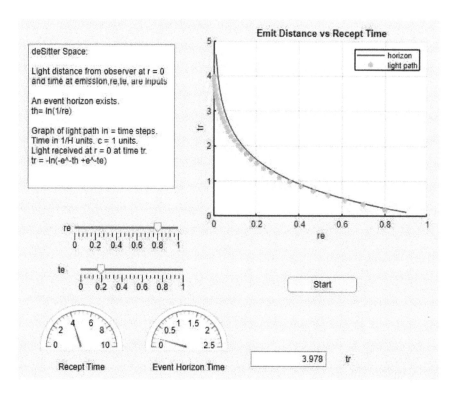

Figure 6.7: Output of the App "deSitterHoriz_App" for a given choice of light emission time and distance from the reception point at $r = 0$.

expands during the light travel time. The horizon time is shown as a continuous blue line. If the light is outside the horizon, the movie is not plotted. The result of a specific choice of parameters is shown in Figure 6.7.

Appendix A

App Scripts

Examples of Matlab scripts are given for the code used in the first chapter of the text. They should serve to give a first Matlab user some experience in fashioning a complete script with both input and output options. Comments appear in green. Character strings and symbolic variables appear as purple text. Numeric variables and executable utilities appear in black. Control of logical operations are shown in blue. Matlab is a vector language, so that operations are often made to all elements of a vector with a single command.

A.1. Taylor Script

```
%
% Symbolic Math - Taylor expansion
%
close all;
clear all;
help SM_Taylor      % Clear the memory and print header
%
% Initialize
%
syms  a b n x Y N yy fin ftay
%
fprintf('Symbolic Math Taylor Expansion: enter f(x) \n ')
fprintf(' An Example function, offset and # terms  \n')
fin = cos(x)
aoff = 0
nterm = 5
ftay = taylor(fin,x,aoff,'Order',nterm);
yy = simplify(ftay);
```

199

```
pretty(yy)
%
figure
xx = linspace(-5,5);
x = xx;
fx = eval(fin);
tfx = eval(ftay);
plot(xx,fx,'r-',xx,tfx,':b')
xlabel('x');
ylabel('f(x), Taylor(f(x))')
title(' Taylor Expansion Around x = 0, 5 Terms')
legend('f(x)','Taylor(f(x))')
%
irun = 1;
%
while irun > 0
    krun = menu('Another Function?','Yes','No');
    if krun == 2
        irun = -1;
        break
    end
    %
    if krun == 1
        figure
        syms  a b n x Y N yy fin ftay
        fin = input(' Enter f(x): ');
        nterm = input(' Enter Number of Terms: ');
        aoff = input('Enter a, Expansion About x = a: ');
        ftay = taylor(sym(fin),x,aoff,'Order',nterm);
        yy = simplify(ftay);
        pretty(yy)
        %
        x = xx;
        fx = eval(fin);
        tfx = eval(ftay);
        plot(xx,fx,'r-',xx,tfx,':b')
        xlabel('x');
        ylabel('f(x), Taylor(f(x))')
        title(' f(x) and Taylor Expansion')
```

```
        legend('f(x)','Taylor(f(x))')
          %
     end
end
%
```

A.2. Taylor Live

```
%
% Symbolic Math - Taylor expansion
%
close all;
%
% Initialize
%
syms   a b n x Y N yy fin ftay
%
% Symbolic Math Taylor Expansion: enter f(x)
% An Example function, offset and # terms
%
fin = cos(x); aoff = 0; nterm = 5;
ftay = taylor(fin,x,aoff,'Order',nterm);
yy = simplify(ftay);
pretty(yy)
  4     2
 x     x
 -- - -- + 1
 24     2
%
figure
xx = linspace(-5,5);
x = xx;
fx = eval(fin);
tfx = eval(ftay);
plot(xx,fx,'r-',xx,tfx,':b')
xlabel('x');
ylabel('cos(x), Taylor(f(x))')
title(' cos(x) Taylor Expansion Around x = 0, 5 Terms')
legend('f(x)','Taylor(f(x))')
```

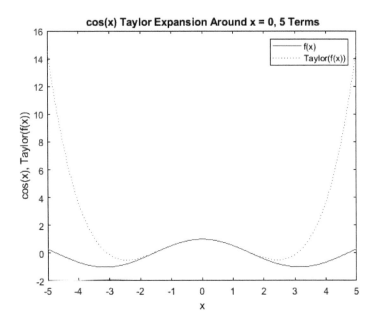

```
%
irun = 1;
%
while irun > 0
    krun = menu('Another Function?','Yes','No');
    if krun == 2
        irun = -1;
        break
    end
    %
    if krun == 1
        figure
        syms  a b n x Y N yy fin ftay
        fin = input(' Enter f(x): ');
        nterm = input(' Enter Number of Terms: ');
        aoff = input('Enter a, Expansion About x = a: ');
        ftay = taylor(sym(fin),x,aoff,'Order',nterm);
        yy = simplify(ftay);
        pretty(yy)
        %
        x = xx;
```

```
fx = eval(fin);
tfx = eval(ftay);
plot(xx,fx,'r-',xx,tfx,':b')
xlabel('x');
ylabel('f(x), Taylor(f(x))')
title(' f(x) and Taylor Expansion')
legend('f(x)','Taylor(f(x))')
%
```

$$x \frac{(2 x^4 + 10 x^2 + 15)\ 2}{15}$$

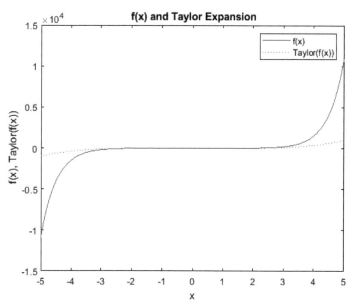

$$\frac{27 x^4}{8} - \frac{9 x^2}{2} + 1$$

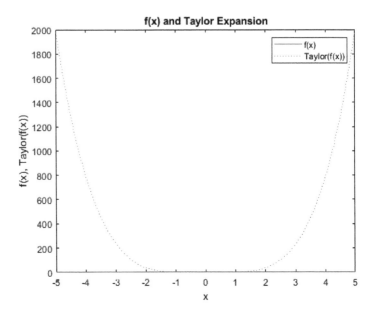

```
        end
    end
    %
```

A.3. Plot_Demo_App

```
methods (Access = private)

    function update(app)
        %
        % plot demo
        %
        x = linspace(-pi,pi);
        omega = app.PhaseFactorSlider.Value;
        y = sin(omega *x);
        plot(app.UIAxes,x,y)
        title(app.UIAxes,'Function sin(phase*\theta)')
        xlabel(app.UIAxes,'\theta')
        ylabel(app.UIAxes,'sin(phase*\theta)')
        grid(app.UIAxes)
```

```
        end
    end

    methods (Access = private)
        % Value changed function: PhaseFactorSlider
        function PhaseFactorSliderValueChanged(app, event)
            value = app.PhaseFactorSlider.Value;
            update(app)
        end
        % Button pushed function: StartButton
        function StartButtonPushed(app, event)
            update(app)
        end
    end
end
```

A.4. 3D Graphics Demo App

```
function updateplot(app)
            %
            colormap(app.UIAxes,'jet')
            ity = app.PlottypeEditField.Value;
            if ity == 1
                [X,Y] = meshgrid(-3:.125:3);
                Z = peaks(X,Y);
                mesh(app.UIAxes,Z);
            end
            %
            if ity == 2
                [X,Y] = meshgrid(-3:.125:3);
                Z = peaks(X,Y);
                meshc(app.UIAxes,Z);
            end
            %
            if ity == 3
                [X,Y,Z] = peaks(30);
                surf(app.UIAxes,X,Y,Z);
                xlim(app.UIAxes, [-3 3]);
                ylim(app.UIAxes, [-3 3]);
                zlim(app.UIAxes, [-10 8]);
```

```
        end
        %
        if ity == 4
            [X,Y,Z] = peaks(30);
            surfc(app.UIAxes,X,Y,Z);
            xlim(app.UIAxes, [-3 3]);
            ylim(app.UIAxes, [-3 3]);
            zlim(app.UIAxes, [-10 8]);
        end
    end
```

A.5. Data Histogram App

```
function update(app)
        %
        % Data Examination
        %
        % Monte Carlo Lorentzian "data"
        %
        a = 10; b = 3; xmi = 0; xmx = 20; % mean a, FWHM b, min, max
        phmi = (2 .*(xmi - a)) ./b;
        phmx = (2 .*(xmx - a)) ./b;
        number = app.numberEditField.Value;
        for i = 1:number
           x(i) = a + (b ./2) .*tan(atan(phmi) + rand .*(atan(phmx) -
           atan(phmi)));
        end
        %
        % histogram the MC "data", prune outliers with x bin limits
        %
        nbin = 50;
        indx = 0;
        %
        % convert event vector x into hist vector h for plots,fits
        %
        % make hist over the range xmi, xmx with nbins - equally spaced.
        % get the appropriate boundaries of the bins - nbin+1 boundaries
        %
        xbound = linspace(xmi,xmx,nbin+1); % lower boundary, equal spacing
        %
        % find bin center as appropriate fit point
```

```
%
xp = (xbound(2)-xbound(1)) ./2 + xbound(1:nbin); % center value,
nbins
h = hist(x,xp); % n.b. first and last bin contains uf and of
e = sqrt(h);   % assume statictical errors
%
ht = sum(h);
app.meanEditField.Value = mean(x);
app.rmsEditField.Value = std(x);
%
%
errorbar(app.UIAxes,xp,h,e,'o')
title(app.UIAxes,'Monte Carlo - Resonance Lorentzian')
xlabel(app.UIAxes,'x')
ylabel(app.UIAxes,'Hist #, Mean = 10, Width = 3')
%
end
```

A.6. Straight Line Fit App

```
function update(app)
    %
    % Program to look at least squares fit to a stright line, variable
      errors
    % least squares fit to a st line y=m*x+b to points y err wt=1/sig^2
      at x
    % data on temp along rod between 2 heat baths at fixed temperatures
    %
    x = [1:9];
    T = [ 15.6 17.5 36.6 43.8 58.2 61.6 64.2 70.4 98.8];
    %
    errscale = app.errorSlider.Value;
    sig = T * errscale ./100; % constant % error
    wt = 1.0 ./(sig .^2);
    [m,b,chi,err] = leastsq(app,x,T,wt);
    dof = length(x) - 2;
    dm = sqrt(err(1));
    db = sqrt(err(4));
    %
    % output the fit
    app.mEditField.Value = m;
    app.dmEditField.Value = dm;
```

```
      app.bEditField.Value = b;
      app.dbEditField.Value = db;
      app.chisqdofEditField.Value = chi ./dof;
      %
      errorbar(app.UIAxes,x,T,sig,'o')
      hold(app.UIAxes, 'on')
      xx= linspace(1,9);
      yy = m .*xx + b;
      plot(app.UIAxes,xx,yy,'-')
      hold (app.UIAxes,'off')
      title(app.UIAxes,'Some Data and Best Straight Line Fit')
      xlabel(app.UIAxes,'Position')
      ylabel(app.UIAxes,'Temperature')

  end

  function[m,b,chi,err] = leastsq(app,x,y,wt)
      %
      % array operations
      %
      xyss = sum(x .* y .*wt);
      xss = sum(x .*wt);
      yss = sum(y .*wt);
      ss = sum(wt);
      xxss = sum(x .*x .*wt);
      %
      d = xxss .*ss - xss .*xss;
      if d == 0.,
        chi= -100.;
      else
        m = (xyss .*ss - xss .*yss) ./d;
        b = (xxss .*yss - xyss .*xss) ./d;
        chi = sum((y-m .*x - b) .^2 .*wt);
        err(1) = ss ./d;
        err(2) = -xss ./d;
        err(3) = err(2);
        err(4) = xxss ./d;
      end

  end
```

A.7. ODE Example App

```
function update(app)
        %
        % ode example
        %
        global a
        %
        a = 1;
        %
        tspan = linspace(0,10,200);
        [tt,yy] = ode45(@app.odedemofun2,tspan,1.0);
        plot(app.UIAxes,tt,yy)
        xlabel(app.UIAxes,'t')
        ylabel(app.UIAxes,'y')
        title(app.UIAxes,'ode demonstration, dy/dt = a*cos(y)*exp(y)')
        %
    end

    function dydt = odedemofun2(app,t,y)
        global a
        %
        dydt = - a .* cos(y)* exp(y);
        %
    end
```

A.8. BVP Example App

```
function update(app)
        %
        % look at solving boundary value problem for diffusion
        % radiation damage
        %
        % guess based on steady state , number of initial x points
        %
        solinit = bvpinit(linspace(0,1,5),@app.Rad_Init2);
        %
        sol = bvp4c(@app.Rad_DE2, @app.Rad_BC2,solinit);
        %
        x = linspace(0,1);
        y = deval(sol,x);
        plot(app.UIAxes,x,y(1,:));
```

```
    xlabel(app.UIAxes,'x/L')
    ylabel(app.UIAxes,'O(x)/O(0)')
    title(app.UIAxes,'Steady State O_2 Distribution in Slab')
    xlim(app.UIAxes,[0 1]);
    ylim(app.UIAxes,[0 1]);
    hold(app.UIAxes,'on');
    %
end

function vinit = Rad_Init2(app,x)
    %
    del = app.aSlider.Value;
    gam = app.bSlider.Value;
    % solve DE if beta*y >> 1
    ylin = (gam ./(2.0 .*del)) .*x .*(x - 1.0) + 1.0;
    dylin = (gam ./del) .*(x - 0.5);
    % solve if betay << 1
    xx = sqrt(gam .*x); gg = sqrt(gam);
    det = exp(gg) - exp(-gg);
    aa = (exp(gg) - 1) ./det; bb = (1 - exp(-gg)) ./det;
    ylin2 = aa .*exp(-xx) + bb .*exp(xx);
    if ylin < 0.1
        ylin = 0.1;
        dylin = 0;
    end
    %vinit = [abs(sin((pi .*x) ./2)), too soft
    %           0];
    % solution when beta*y >> 1
    vinit = [ylin
            %0];
            dylin];
    % solution when beta*y << 1
    %vinit = [ylin2
    %           0];
end

function res = Rad_BC2(app,ya,yb)
    % normalize to 1 at the tile boundaries
    % residuals
    res = [ya(1) - 1 ; yb(1) - 1];
end

function dydx = Rad_DE2(app,x,y)
```

```
        del = app.aSlider.Value;
        gam = app.bSlider.Value;
        dydx = [y(2) ; gam .*y(1) ./(1.0 + del .*y(1))];
    end
```

A.9. Taylor_App

```
function update(app)
        %
        % Symbolic Math - Taylor expansion
        %
        syms  a b n x Y N yy zz fin ftay
        %
        % Symbolic Math Taylor Expansion: enter f(x)
        %   offset and # terms
        %
        itay = app.FunctionEditField.Value;%
        %
        if itay == 1
            fin = cos(x);
        end
        if itay == 2
            fin = sin(x);
        end
        if itay == 3
            fin = cosh(x);
        end
        if itay == 4
            fin = sinh(x);
        end
        if itay == 5
            fin = exp(x);
        end
        aoff  = app.expandaroundEditField.Value;
        nterm = app.numberoftermsEditField.Value;
        ftay = taylor(fin,x,aoff,'Order',nterm);
        yy = simplify(ftay);
        pretty(yy)
        app.seriesEditField.Value = char(yy);
        %
        xlim = app.SymmetriclimitEditField.Value;
        xx = linspace(-xlim,xlim);
```

```
        x = xx;
        fx = eval(fin);
        tfx = eval(ftay);
        plot(app.UIAxes,xx,fx,'r-',xx,tfx,':b')
        legend(app.UIAxes,'function','series')
    end
```

Appendix B

Symbols

a semi major axis of ellipse, acceleration, Kerr rotation parameter

A Angstrom, atomic weight, Area

b semi minor axis of ellipse

B_A binding energy

B magnetic field

c speed of light

c_s speed of sound

C capacitance

D deuterium

ds space-time interval

$d\Omega$ differential solid angle

e orbital eccentricity, electron, electron charge

eV electron volt $= 1.6 \times 10^{-19}$ J

E electric field

e^- electron

f frequency, flux

F force, energy flux (W/m^2)

F_H heat flux (W/m^2)

g acceleration of gravity on Earth

G gravitational constant

G_F Fermi weak decay constant

h Planck constant, metric distortion scale, tidal height. SR constant of motion, approximately J/mc

$\hbar = h/2\pi$, reduced Planck constant

H Hubble constant

I electric current

J angular momentum

j current per area

k Boltzmann constant

K kinetic energy

k_λ wave number

L luminosity

M mass of a body, Mach number

m mass of a smaller body or a particle

m_p payload mass

n, p neutron, proton

n number density

N total number of objects

NSE Nuclear Statistical Equilibrium

p particle momentum, proton

P pressure

P_c core pressure

ψ wave function

ω circular frequency

μ mean molecular weight factor, 2 body reduced mass

μ_o vacuum permeability

Q n p mass difference, charge

q particle electric charge

R radius of a body, gas constant, electrical resistance

RR reaction rate

R_s Schwarzschild radius

s proper time

S stress, tensile strength

T temperature, maximum rocket burn time, Tritium

T_c core temperature

u energy density, $1/r$ for orbits

U total energy

V volume, potential energy

v neutrino

v_o rocket exhaust velocity

v_s speed of sound

W power

X mass fraction w.r.t. nucleons

Z atomic number

β velocity$/c$

γ photon, ratio of specific heats, SR time dilation factor

Γ inverse mean free path, reaction rate, Lorentzian FWHM

Δ grid spacing

χ the chisquared value of a fit

ε particle energy, solar energy production

ε_o vacuum permitivity

η nucleon to photon ratio

κ opacity

λ wavelength

μ mean molecular weight, ratio of specific heats, reduced mass

μ_o vacuum permeability

ρ mass or charge density

ρ_c core density

σ Stefan–Boltzmann constant, Gaussian standard deviation, cross section

θ polar angle, polytropic function

τ period

ϕ azimuthal angle

Φ potential per charge or mass

Appendix C

Properties of the Sun

Property	Symbol	Value
Radius	$R(\mathrm{m})$	6.9×10^8
Mass	$M(\mathrm{kg})$	2.0×10^{30}
Luminosity	$L(\mathrm{W})$	3.8×10^{26}
Surface temperature	$T(^\circ\mathrm{K})$	5770
Solar constant (Earth)	$f(\mathrm{W/m^2})$	1.37×10^3
Schwarzschild radius	$R_s(\mathrm{km})$	3.0
Number density $-$ mean	$n(1/\mathrm{m^3})$	1.2×10^{23}
Mass density $-$ mean	$\rho(\mathrm{kg/m^3})$	1.4×10^3
Pressure $= GM\rho/R$ $-$ mean	$P(\mathrm{Pa} = \mathrm{Nt/m^2})$	3×10^{14}
Core temperature $= T_c$	$T(^\circ\mathrm{K})$	1.5×10^7
Core pressure $= P_c$	$P(\mathrm{Pa})$	2.6×10^{16}
Core density $= \rho_c$	$\rho_c(\mathrm{kg/m^3})$	1.5×10^5
Hydrogen and helium mass fraction, X and Y. $X + Y \sim 1$	X, Y	0.7, 0.28
X and Y, core values, ionized	X, Y	0.35, 0.65
Mean molecular mass fraction, ionized	μ	1.30 H, 0.62

Appendix D

Properties of the Planets

Planet	M/M_{Earth}	R/R_{Earth}	$a(AU)$	$\tau(\mathrm{yr})$
Mercury	0.055	0.38	0.39	0.24
Venus	0.81	0.95	0.72	0.61
Earth	1.0	1.0	1.0	1.0
Mars	0.11	0.53	1.52	1.9
Jupiter	318	11.2	5.2	11.9
Saturn	95	9.4	9.6	29.5
Uranus	14.5	4.0	19.2	84.0
Neptune	17.1	3.9	30.0	165

Appendix E

Properties of the Earth's Atmosphere

$PV = NkT$

$\langle \varepsilon \rangle = (3/2)kT, \; PV = (2/3)\langle \varepsilon \rangle$

Property	Value
Density (kg/m^3)	1.3
Pressure (bar, $Pa = $Nt/m^2,Torr)	1, 10^5, 760
Sound speed (km/sec) at STP	$\sim$0.32

Appendix F

Acronyms

BVP	Boundary Value Problem
CM	Classical Mechanics or Center of Mass of a system
CMB	Cosmic Microwave Background
EM	Electromagnetism
ESA	European Space Agency
FWHM	Full Width at Half Maximum
GEO	Geosynchronous Earth Orbit
GR	General Relativity
ISM	Inter-Stellar Medium
ISS	International Space Station
ITER	International Thermonuclear Experimental Reactor
LEO	Low Earth Orbit
LIGO	Laser Interferometer Gravity-wave Observatory
LISA	Laser Interferometer Space Antenna
NASA	National Aeronautics and Space Administration
NR	Non-Relativistic
NSE	Nuclear Statistical Equilibrium
ODE	Ordinary Differential Equation
PDE	Partial Differential Equation
QM	Quantum Mechanics
SHM	Simple Harmonic Motion
SR	Special Relativity
STP	Standard Temperature and Pressure
UR	Ultra Relativistic

References

It is traditional to make a long list of references in print textbooks and articles. However, in the present state of the internet with many documents digitized and with many useful search engines, it seems appropriate to simply use these tools. An example of the first few results of a query for "physics of a relativistic rocket" appears below. There are links to scholarly journals, many of which are now open source, publicly available. These tools are therefore, now becoming freely available to a worldwide readership. Wikipedia is a good starting point for general queries.

Recently there has been a growing movement to make scientific preprints and journals open source. The argument is that as research is often supported by government agencies, the results of that research should be freely available to the people. An example for preprints in the field of math and science is the "arXiv" site. This site has many sub-sites which cover preprints of current research. Indeed, these preprints often are more timely than the actually published papers whose time for publication can be rather long. A page of arXiv for "Physics" is shown in Figure R.2. There are several other subsets, including Mathematics and Computer Science and the Wikipedia result for a search on "arXiv" appears in Figure R.3.

Relativistic rocket - Wikipedia
https://en.wikipedia.org/wiki/Relativistic_rocket

         ⑦ About this result   🎟 Feedback

The Relativistic Rocket - UCR Math
math.ucr.edu/home/baez/physics/Relativity/SR/Rocket/rocket.html ▾
[**Physics** FAQ] - [Copyright]. Thanks to Aron Yoffe for suggesting the newtonian comparison. Thanks to
Tom Fuchs for suggesting the analogy as the stone thrown ...
You've visited this page 2 times. Last visit: 7/17/19

The Relativistic Rocket: American Journal of Physics: Vol 34, No 7
https://aapt.scitation.org/doi/10.1119/1.1973114
The equation of motion for a **relativistic rocket** or for any particle with a variable rest mass may be
derived by applying the relativistic impulse-momentum and ...

Relativistic Rocket Theory: American Journal of Physics: Vol 21, No 4
https://aapt.scitation.org/doi/10.1119/1.1933430
In the **relativistic** theory of one-dimensional **rocket** motion in empty space, ... American Journal of
Physics 21, 310 (1953); https://doi.org/10.1119/1.1933430.
You've visited this page 2 times. Last visit: 7/17/19

Figure R.1:   Results of a search for "physics of a relativistic rocket"

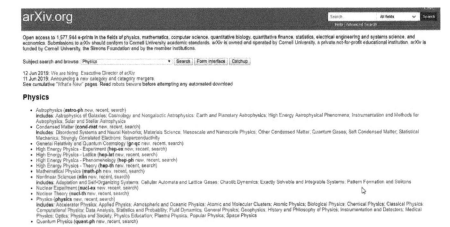

Figure R.2:   Introductory page after a search for "arXiv" and displaying only at
the "Physics" subset of sites.

arXiv

arXiv (pronounced "archive"—the X represents the Greek letter chi [χ])[2] is a repository of electronic preprints (known as e-prints) approved for posting after moderation, but not full peer review. It consists of scientific papers in the fields of mathematics, physics, astronomy, electrical engineering, computer science, quantitative biology, statistics, mathematical finance and economics, which can be accessed online. In many fields of mathematics and physics, almost all scientific papers are self-archived on the arXiv repository. Begun on August 14, 1991, arXiv.org passed the half-million-article milestone on October 3, 2008,[3][4] and had hit a million by the end of 2014.[5][6] By October 2016 the submission rate had grown to more than 10,000 per month.[6][7]

Figure R.3:   Beginning of a long text on the free arXiv site for scientific preprints.

https://alfven.princeton.edu/publications/ep-encyclopedia-2001

An example of a paper from a University physics department in regards to ion propulsion appears below. It seems clear that there are abundant resources now available world-wide on a free and open basis. The need for expensive texts often places them beyond the means of young students, and the internet and open source publication helps to remove those obstacles. The site for access to the paper with the heading text is shown below in Figure R.4.

Electric Propulsion

Robert G. Jahn
Edgar Y. Choueiri
Princeton University

   I. Conceptual Organization and History of the Field
  II. Electrothermal Propulsion
 III. Electrostatic Propulsion
 IV. Electromagnetic Propulsion
  V. Systems Considerations
 VI. Applications

Figure R.4:   Text of the first page of a specific paper discussing ion propulsion.

Index

Lightning Source UK Ltd.
Milton Keynes UK
UKHW020905240520
363673UK00005B/27